CRC SERIES IN RADIOTRACERS IN BIOLOGY AND MEDICINE

Editor-in-Chief

Lelio G. Colombetti, Sc.D.
Loyola University
Stritch School of Medicine
Maywood, Illinois

STUDIES OF CELLULAR FUNCTION USING RADIOTRACERS
Mervyn W. Billinghurst, Ph.D.
  Radiopharmacy
  Health Sciences Center
  Winnipeg, Manitoba, Canada

GENERAL PROCESSES OF RADIOTRACER LOCALIZATION
Leopold J. Anghileri, D.Sc.
  Laboratory of Biophysics
  University of Nancy
  Nancy, France

RADIATION BIOLOGY
Donald Pizzarello, Ph.D.
  Department of Radiology
  New York University Medical Center
  New York, New York

RADIOTRACERS FOR MEDICAL APPLICATIONS
Garimella V. S. Rayudu, Ph.D.
  Nuclear Medicine Department
  Rush University Medical Center
  Presbyterian-St. Luke's Hospital
  Chicago, Illinois

RECEPTOR-BINDING RADIOTRACERS
William C. Eckelman, Ph.D.
  Department of Radiology
  George Washington University School of Medicine
  Washington, D.C.

BIOLOGIC APPLICATIONS OF RADIOTRACERS
Howard J. Glenn, Ph.D.
  University of Texas System Cancer Center
  M.D. Anderson Hospital and Tumor Institute
  Houston, Texas

BIOLOGICAL TRANSPORT OF RADIOTRACERS
Lelio G. Colombetti, Sc.D.
  Loyola University
  Stritch School of Medicine
  Maywood, Illinois

BASIC PHYSICS OF RADIOTRACERS
W. Earl Barnes, Ph.D.
  Nuclear Medicine Service
  Edward Hines, Jr., Hospital
  Hines, Illinois

RADIOBIOASSAYS
Fuad S. Ashkar, M.D.
  Radioassay Laboratory
  Jackson Memorial Medical Center
  University of Miami School of Medicine
  Miami, Florida

COMPARTMENTAL DISTRIBUTION OF RADIOTRACERS
James S. Robertson, M.D., Ph.D.
  Mayo Medical School
  Mayo Clinic
  Rochester, Minnesota

RADIONUCLIDES PRODUCTION
Frank Helus, Sc.D.
  Institute of Nuclear Medicine
  German Cancer Research Center
  Heidelberg, Germany

Radionuclides Production

Volume II

Editor

Frank Helus, Ph.D.
Institute of Nuclear Medicine
German Cancer Research Center
Heidelberg, West Germany

Editor-in-Chief
CRC Series in Radiotracers in Biology
and Medicine

Lelio G. Colombetti, Ph.D.
Loyola University
Stritch School of Medicine
Maywood, Illinois

CRC Press, Inc.
Boca Raton, Florida

Library of Congress Cataloging in Publication Data
Main entry under title:

Radionuclides production.

(CRC series in radiotracers in biology and
medicine
Bibliography: p.
Includes index.
1. Radioisotopes. I. Helus, Frank. II. Series.
[DNLM: 1. Radioisotopes. 2. Nuclear physics--
Methods.   QC 795.7 R129]
QD601.2.R35   1983   621.48'37   83-2827
ISBN 0-8493-6003-X (v. 1)
ISBN 0-8493-6004-8 (v. 2)

This book represents information obtained from authentic and highly regarded sources. Reprinted material is quoted with permission, and sources are indicated. A wide variety of references are listed. Every reasonable effort has been made to give reliable data and information, but the author and the publisher cannot assume responsibility for the validity of all materials or for the consequences of their use.

All rights reserved. This book, or any parts thereof, may not be reproduced in any form without written consent from the publisher.

Direct all inquiries to CRC Press, Inc., 2000 Corporate Blvd., N.W., Boca Raton, Florida, 33431.

© 1983 by CRC Press, Inc.

International Standard Book Number 0-8493-6003-X (Vol.1)
International Standard Book Number 0-8493-6004-8 (Vol.2)

Library of Congress Card Number 83-2827
Printed in the United States

FOREWORD

This series of books on Radiotracers in Biology and Medicine is on the one hand an unbelievably expansive enterprise and on the other hand, a most noble one as well. Tools to probe biology have developed at an accelerating rate. Hevesy pioneered the application of radioisotopes to the study of chemical processes, and since that time, radioisotopic methodology has probably contributed as much as any other methodology to the analysis of the fine structure of biologic systems. Radioisotopic methodologies represent powerful tools for the determination of virtually any process of biologic interest. It should not be surprising, therefore, that any effort to encompass all aspects of radiotracer methodology is both desirable in the extreme and doomed to at least some degree of inherent failure. The current series is assuredly a success relative to the breadth of topics which range from in depth treatise of fundamental science or abstract concepts to detailed and specific applications, such as those in medicine or even to the extreme of the methodology for sacrifice of animals as part of a radiotracer distribution study. The list of contributors is as impressive as is the task, so that one can be optimistic that the endeavor is likely to be as successful as efforts of this type can be expected to be. The prospects are further enhanced by the unbounded energy of the coordinaing editor. The profligate expansion of application of radioisotopic methods relate to their inherent and exquisite sensitivity, ease of quantitation, specificity, and comparative simplicity, especially with modern instrumentation and reagents, both of which are now readily and universally available. It is now possible to make biological measurements which were otherwise difficult or impossible. These measurements allow us to begin to understand processes in depth in their unaltered state so that radioisotope methodology has proved to be a powerful probe for insight into the function and perturbations of the fine structure of biologic systems. Radioisotopic methodology has provided virtually all of the information now known about the physiology and pathophysiology of several organ systems and has been used abundantly for the development of information on every organ system and kinetic pathway in the plant and animal kingdoms. We all instinctively turn to the thyroid gland and its homeostatic interrelationships as an example, and an early one at that, of the use of radioactive tracers to elaborate normal and abnormal physiology and biochemistry, but this is but one of many suitable examples. Nor is the thyroid unique in the appreciation that a very major and important residua of diagnostic and therapeutic methods of clinical importance result from an even larger number of procedures used earlier for investigative purposes and, in some instances, procedures used earlier for investigative purposes and, in some instances, advocated for clinical use. The very ease and power of radioisotopic methodology tempts one to use these techniques without sufficient knowledge, preparation or care and with the potential for resulting disastrous misinformation. There are notable research and clinical illustrations of this problem, which serve to emphasize the importance of texts such as these to which one can turn for guidance in the proper use of these powerful methods. Radioisotopic methodology has already demonstrated its potential for opening new vistas in science and medicine. This series of texts, extensive though they be, yet must be incomplete in some respects. Multiple authorship always entails the danger of nonuniformity of quality, but the quality of authorship herein assembled makes this likely to be minimal. In any event, this series undoubtedly will serve an important role in the continued application of radioisotopic methodology to the exciting and unending, yet answerable, questions in science and medicine!

Gerald L. DeNardo, M.D.
Professor of Radiology, Medicine,
Pathology and Veterinary Radiology
University of California, Davis-
Sacramento Medical School
Director, Division of Nuclear Medicine

THE EDITOR-IN-CHIEF

Lelio G. Colombetti, Sc.D., is Professor of Pharmacology at Loyola University Stritch School of Medicine in Maywood, Ill. and a member of the Nuclear Medicine Division Staff at Michael Reese Hospital and Medical Center in Chicago, Ill.

Dr. Colombetti graduated from the Litoral University in his native Argentina with a Doctor in Sciences degree (summa cum laude), and obtained two fellowships for postgraduate studies from the Georgetown University in Washington, D.C., and from the M.I.T. in Cambridge, Mass. He has published more than 150 scientific papers and is the author of several book chapters. He has presented over 300 lectures both at meetings held in the U.S. and abroad. He organized the First International Symposium on Radiopharmacology, held in Innsbruck, Austria, in May 1978. He also organized the Second International Symposium on Radiopharmacology which took place in Chicago in September, 1981, with the active participation of more than 500 scientists, representing over 30 countries. He is a founding member of the International Association of Radiopharmacology, a nonprofit organization, which congregates scientists from many disciplines interested in the biological applications of radiotracers. He was its first President (1979/1981.

Dr. Colombetti is a member of various scientific societies, including the Society of Nuclear Medicine (U.S.) and the Gesellschaft für Nuklearmedizin (Europe), and is an honorary member of the Mexican Society of Nuclear Medicine. He is also a member of the Society of Experimental Medicine and Biology, the Coblenz Society, and the Sigma Xi. He is a member of the editorial boards of the journals *Nuklearmedizin* and *Research in Clinic and Laboratory*.

THE EDITOR

Frank Helus, Sc.D., is leading the radiochemistry group and is the person responsible for the production of cyclotron and reactor radionuclides at the Institute of Nuclear Medicine, German Cancer Research Center, Heidelberg. He studied nuclear chemistry and radiochemistry at Prague Charles University and the Technical University of Prague. He received his degree in nuclear chemistry from Technical University of Prague in 1960. After his studies he joined the academic staff at the Department of Nuclear Chemistry at the Technical University of Prague and conducted both theoretical (hot atom chemistry lecture) and practical classes for students. At the same time he made extensive research work in the field of the chemical behavior of the trace products after nuclear reactions. He moved to Italy in 1967, where he worked on the preparation of technetium labeled radiopharmaceuticals for medical use, and in 1969 he joined the Hammersmith Cyclotron Unit M.R.C. research group in London and began working on the production of cyclotron radionuclides. In 1971 he moved to Heidelberg where he is now in charge of the Radiochemical group with specific interests in the production and labeling processes with short-lived positron emitters and non-invasive physiological studies performed through the use of such nuclides.

Frank Helus is known for his research on the production of Br-77 and F-18 and is authority in the production of cyclotron radionuclides for medical use; more specifically in cyclotron targetry and fast separation procedures. His publications comprise more than 50 research papers. He is a member of various scientific societies.

CONTRIBUTORS

Zeev Alfassi, Ph.D.
Associate Professor
Nuclear Engineering Department
Ben Gurion University
Beer Sheva, Israel

R. E. Boyd, Ph.D.
Head, Radioactive Products
Research Section, Isotope Division
Australian Atomic Energy Commission
Research Establishment
Lucas Heights Research Laboratories
Sutherland, Australia

Frank Helus, Sc.D.
Institute of Nuclear Medicine
German Cancer Research Center
Heidelberg, West Germany

R. S. Mani, Ph.D.
Head
Radiopharmaceutical Section
Isotope Group
Bhabha Atomic Research Center
Bombay, India

Milorad Mladjenovic, Ph.D.
Professor
Boris Kidric Institute
University of Belgrade
Belgrade, Yugoslavia

Tadashi Nozaki, Ph.D.
Principal Scientist
Radiochemistry Laboratory
Rikagaku Kenkyu-sho
Saitama, Japan

A. M. J. Paans, Ph.D
Department of Nuclear Medicine
University Hospital
Groningen, The Netherlands

G. D. Robinson, Jr., Ph.D.
Research Associate Professor of Neurology and Radiology
Director, Medical Cyclotron Facility
University of Pennylvania
Philadelphia, Pennsylvania

E. L. Sattler, Ph.D.
Strahlenzentrum
Giessen, West Germany

David J. Silvester, Ph.D.
Head, Chemistry Section
Medical Research Council Cyclotron Unit
Hammersmith Hospital
London, England

W. Vaalburg, Ph.D.
Head, Cyclotron Radiopharmaceutical Group
Department of Nuclear Medicine
University Hospital
Groningen, The Netherlands

Gerd Wolber, Ph.D.
Head, Cyclotron Group
German Cancer Research Center
Heidelberg, West Germany

TABLE OF CONTENTS

Volume I

Chapter 1
Radioisotope Production: An Historical Introduction 1
D. J. Silvester

Chapter 2
Nuclear Physics Fundamentals 11
Milorad Mladjenovic

Chapter 3
Activation Techniques 57
Frank Helus and G. Wolber

Chapter 4
Radiochemical Processing of Activated Targets 121
G. D. Robinson, Jr.

Index 139

Volume II

Chapter 1
Reactor-Produced Radionuclides 1
R. S. Mani

Chapter 2
Short-Lived Positron Emitting Radionuclides 47
W. Vaalburg and A. M. J. Paans

Chapter 3
Other Cyclotron Radionuclides 103
Tadashi Nozaki

Chapter 4
The Special Position of $^{99}\Sigma$Tc in Nuclear Medicine 125
R. E. Boyd

Chapter 5
Production of Radionuclides by 14 MeV Neutron Generator 153
Zeev B. Alfassi

Chapter 6
Radionuclides and Labeled Compounds Produced at an Electron Linear Accelerator ... 161
E. L. Sattler

Index 169

Chapter 1

REACTOR-PRODUCED RADIONUCLIDES

R. S. Mani

TABLE OF CONTENTS

I. Introduction 2

II. General Requirements for Reactor-Produced Radionuclides Used as
 Tracers 3
 A. Practical Half-Life 3
 B. Low Toxicity 5
 C. Purity 6
  1. Radionuclidic Purity 6
  2. Radiochemical Purity 8
  3. Chemical Purity 9
 D. High Specific Activity 9
 E. Ready Availability and Low Price 10
 F. Special Requirements for Preparation of Labeled Compounds and
  Radiopharmaceuticals 11

III. Main Production Reactions 13
 A. (n,γ) Reactions 13
 B. (n,γ) Followed by Decay 14
 C. (n,p) and (n,α) Reactions 16
 D. Fission 18
 E. Successive Neutron Capture Reactions 20
 F. Indirect Reactions 20
 G. Szilard-Chalmers Recoil Enrichment 20

IV. Calculation of Yields and Specific Activities 21

V. Large-Scale Production Considerations 24
 A. Irradiation Facilities 24
 B. Raw Materials Including Targets 24
 C. Handling and Processing Facilities 25
  1. Ventilation System 25
  2. Air Conditioning 25
  3. Remote Handling Facilities 25
  4. Effluent, Waste Collection, and Storage 26
  5. Control of Radiation Hazards 26
 D. Treatment and Disposal of Wastes 26
 E. Sales and Marketing 26
 F. Other Requirements for Large-Scale Production 26

VI. Typical Production Processes 26
 A. Tritium 27
 B. Carbon-14 27
 C. Sodium-24 27

	D.	Potassium-42 and Rubidium-86	27
	E.	Phosphorus-32	28
	F.	Sulfur-35	28
	G.	Chromium-51	28
		1. (n,γ) Activation of Enriched ^{50}Cr	28
		2. Szilard-Chalmers Enrichment	28
	H.	Cobalt-58	29
	I.	Iron-59	29
	J.	Molybdenum-99	29
		1. Direct Irradiation of MoO_3	29
		2. Fission of Uranium	29
	K.	Iodine-125	30
	L.	Iodine-131	30
		1. Wet Distillation Methods	30
		2. Dry Distillation Method	31
		3. Relative Advantages and Disadvantages	31
VII.	Quality Controls		31
	A.	Evaluation of Radionuclidic Purity	31
	B.	Evaluation of Radiochemical Purity	32
	C.	Evaluation of Chemical Purity	32
	D.	Other Aspects	32
VIII.	Packaging and Transport		32
	A.	IAEA Regulation of 1967	33
	B.	1973 IAEA Regulations	34
IX.	Current Availability of Reactor-Produced Radionuclides		35
X.	Future Perspectives		35
XI.	Conclusion		38
XII.	Physical Data for Important Reactor-Produced Radionuclides of Interest as Tracers in Biology and Medicine		41
References			44

I. INTRODUCTION

The nuclear reactor is the main source of the large majority of radionuclides which are presently employed as tracers in biology and medicine. Over 50 radionuclides in various chemical forms are currently of interest as tracers in the biomedical sciences, the major part of these applications is, however, accounted by 24 radionuclides listed in Table 1; of these 14 are produced exclusively in nuclear reactors, 8 exclusively in cyclotrons, and the remaining 2 can be produced by either means.

The first nuclear chain reactor was constructed and operated successfully in the U.S. in 1942, and production of radionuclides using this reactor was commenced at the Oak Ridge National Laboratory in 1946. It is of interest to recall that prior to the operation of the first nuclear reactor, several important tracer studies had been carried out using a small spectrum

Table 1
IMPORTANT RADIONUCLIDES USED AS TRACERS IN BIOLOGY AND MEDICINE

Radionuclide	Source for large-scale production
^{3}H (Tritium)	Reactor
^{11}C	Cyclotron
^{13}N	Cyclotron
^{14}C	Reactor
^{15}O	Cyclotron
^{18}F	Reactor, cyclotron
^{24}Na	Reactor
^{35}S	Reactor
^{32}P	Reactor
^{51}Cr	Reactor
^{57}Co	Cyclotron
^{58}Co	Reactor
^{59}Fe	Reactor
^{67}Ga	Cyclotron
^{82}Br	Reactor
^{85}Sr	Reactor, cyclotron
^{99}Mo-^{99m}Tc	Reactor
^{111}In	Cyclotron
^{113}Sn-^{113m}In	Reactor
^{123}I	Cyclotron
^{125}I	Reactor
^{131}I	Reactor
^{169}Yb	Reactor
^{201}Tl	Cyclotron

of naturally occurring and cyclotron-produced radionuclides which were available in small quantities during the period 1913 to 1945. A brief historical survey of important developments in the field of tracer applications of radionuclides during the period 1913 to 1945 is given in Table 2.

With the construction and operation of many nuclear reactors in the U.S., U.K., Europe, and elsewhere in the post second world war years, regular production and supply of a wide variety of radionuclides in various chemical forms became a well-organized industry. Side by side came the development of advanced instrumentation such as liquid scintillation counters, rectilinear scanners, gamma cameras with on-line computer, whole-body counters, and automated sample counting and data processing systems for radioanalytical procedures such as radioimmunoassays. These developments, in turn, provided a great fillip for the growth of a multitude of ingenious applications of radionuclides in biology and medicine.

II. GENERAL REQUIREMENTS FOR REACTOR-PRODUCED RADIONUCLIDES USED AS TRACERS

Reactor-produced radionuclides meant for use as tracers in biology and medicine should have the following criteria and requirements.

A. Practical Half-Life

The physical half-life of the radionuclide should be in the range of a few minutes to a few years. Too short a half-life will pose problems of very high decay losses during preparation, transport, and actual use. Some of the ultra short-lived radionuclides employed for tracer studies and their characteristics are shown in Table 3. Radionuclides having half-lives exceeding several years are not quite suitable for tracer applications for three main reasons,

Table 2
HISTORICAL SURVEY OF DEVELOPMENTS IN TRACER APPLICATIONS

Year	Development	Ref.
1913	Use of radio elements as tracers in chemistry and physics	Hevesy, G., *Chem. News*, 108, 166, 1913.
1913	Proposal of the name "isotopes"	Soddy, F.,
1923	Use of G. M. counter for tracer applications	Hevesy, G., *Biochem. J.*, 17, 439, 1923.
1924	Use of tracers of lead	Hevesy, G., *C. R. Acad. Sci.*, 179, 291, 1924.
1930	Tracer studies	Hevesy, G. and Wagner, O. H., *Arch. Exp. Pathol. Pharmacol.*, 149, 336, 1930.
1931	Development of the cyclotron	Lawrence, E. O. and Livingston, M. S., *Phys. Rev.*, 38, 834, 1931.
1932	Discovery of the neutron	Chadwick, J., *Nature (London)*, 129, 312, 1932.
1934	Artificial production of radionuclide	Curie, I. and Joliot, F., *Nature (London)*, 133, 201, 1934.
1934	Radioactivity by neutron bombardment; neutron capture reactions	Livingston, M. S. and Henderson, M. C., *Proc. Natl. Acad. Sci. USA*, 20, 470, 1934.
1934	Discovery of the Szilard-Chalmers process for preparation of high specific activity nuclides	Szilard, L. and Chalmers, T. A., *Nature (London)*, 134, 462, 1934.
1937	Clinical use of radioactive sodium	Hamilton, J. G. and Stone, R., *Radiology*, 28, 178, 1937.
1938	Radioactive I in the study of thyroid physiology	Hertz, S., Roberts, A., and Evans, R. D., *Proc. Soc. Exp. Biol. Med.*, 38, 510, 1938.
1938	Discovery of fission of U	Hahn, O. and Strassmann, F., *Naturwissenschaften*. 26, 755, 1938.
1938	Absorption of radioactive nuclides of Na, K, Cl, Br, and I in normal human subjects	Hamilton, J. G., *Am. J. Physiol.*, 124, 667, 1938.
1939	Use of radioactive P for studies of leukemia	Lawrence, J. H., Scott, K. G., and Tuttle, L. W., *Int. Clin.*, 3, 33, 1939.
1940	Studies of iodine metabolism by thyroid and use of radioiodine in various types of goiter.	Hamilton, J. G. and Soley, M. H., *Am. J. Physiol.*, 131, 135, 1940.
1940	Discovery of Carbon-14	Ruben, S. and Kamen, M. D., *Phys. Rev.*, 57, 549, 1940.
1941	Applications of radiotracers in biology and medicine	Hamilton, J. G., *J. Appl. Phys.*, 12, 440, 1941.
1941—42	The concept of dynamic state of body constituents	Schoenheimer, R., *The Dynamic State of Body Constituents*, Harvard University Press, Cambridge, Mass., 1946.
1942	Radioactive iodine in the therapy of Grave's Disease	Hertz, S. and Roberts, A., *J. Clin. Invest.*, 21, 31, 1942.
1942	Therapeutic applications of radiophosphorus and radioiodine	Hamilton, J. G., Lawrence, J. H., *J. Clin. Invest.*, 21, 624, 1942.
1942	Construction and operation of the first nuclear reactor	E. Fermi and Associates, University of Chicago
1945—46	Therapy with radioactive colloids by various routes of injection	Hahn, P. F. and Sheppard, C. W, *South. Med. J.*, 39, 558, 1946; *J. Lab. Clin. Med.*, 32, 1437, 1947.

Table 3
ULTRA SHORT-LIVED REACTOR-PRODUCED NUCLIDES OF INTEREST FOR TRACER APPLICATIONS

Radionuclide	Half-life	Reaction for production
^{18}F	109.7 m	^{6}Li (n, α) ^{3}H (2.73 MeV tritons) ^{16}O (^{3}H, n) ^{18}F
^{28}Al	2.25 m	i) ^{27}Al (n, γ) ^{28}Al (Low specific activity) ii) Daughter of ^{28}Mg $\xrightarrow{\beta^-}$ ^{28}Al
^{38}Cl	37.29 m	i) ^{37}Cl (n, γ) ^{38}Cl (Low specific activity) ii) Daughter of ^{38}S $\xrightarrow{\beta^-}$ ^{38}Ci)
^{66}Cu	5.1 m	^{65}Cu (n, γ) ^{66}Cu
^{69}Zn	55.6 m	i) ^{68}Zn (n, γ) ^{69}Zn ii) Szilard-Chalmers reaction on zinc phthalocyanine
^{70}Ga	21.1 m	^{69}Ga (n, γ) ^{70}Ga
^{80}Br	17.68 m	^{79}Br (n, γ) ^{80}Br
^{87}Kr	76.31 m	^{86}Kr (n, γ) ^{87}Kr
^{85m}Sr	68.0 m	^{84}Sr (n, γ) ^{85m}Sr
^{91m}Y	49.71 m	Fission of ^{235}U
^{109m}Pd	4.69 m	^{108}Pd (n, γ) ^{109m}Pd
^{111m}Cd	48.6 m	^{110}Cd (n, γ) ^{111m}Cd
^{113m}In	99.47 m	Daughter of ^{113}Sn which is produced by the ^{112}Sn (n, γ) ^{113}Sn reaction
^{137m}Ba	2.55 m	Daughter of ^{137}Cs [^{235}U (n, f) ^{137}Cs]
^{188m}Re	18.7 m	^{187}Re (n, γ) ^{188m}Re
^{197m}Pt	94.4 m	^{196}Pt (n, γ) ^{197m}Pt
^{199}Pt	30.8 m	^{198}Pt (n, γ) ^{199}Pt

namely, (1) their specific activity, and consequently the number of disintegrations per unit mass, will be too low, (2) there are limitations on the mass of the tracer material which can be employed or administered, and (3) in diagnostic tracer studies in medicine the patient is likely to receive excessively high radiation exposures if long-lived radionuclides are employed. To minimize the patient exposure, short-lived radionuclides have replaced many of their long-lived counterparts which were in use earlier; an example is technetium-99m, having a half-life of 6 hr, which is now preferred to mercury-203 and iodine-131, having half-lives of 46.6 days and 8.04 days, respectively, for in vivo nuclear medicine procedures.

B. Low Toxicity

For tracer applications in biology and medicine, the radionuclide chosen should be of low to medium toxicity. The toxicity of a radionuclide is its ability to produce damage or injury by virtue of its emitted radiations when incorporated in a body. The toxicity depends on the absorption of the nuclide in the body, the manner and sites of deposition, rate of elimination, half-life, and types of radiations emitted. Radionuclides of metabolically important elements such as P, Ca, C, S, and Fe, participate in the metabolic processes and are retained at different sites in the body for long periods of time. Alkaline earth elements like Sr and Ra are metabolized like Ca and are deposited and retained in the bones. Radionuclides of I are selectively taken up by the thyroid gland. The radiations emitted by radionuclides which are thus deposited in specific sites or organs — which are designated as "Critical Organs" — cause considerable damage. These radionuclides are therefore classified among the more

Table 4
CLASSIFICATION OF REACTOR-PRODUCED RADIONUCLIDES OF INTEREST AS BIOMEDICAL TRACERS ACCORDING TO RELATIVE RADIOTOXICITY

Group 1	None
Group 2	^{36}Cl, ^{45}Ca, ^{91}Y, ^{131}I, ^{137}Cs, ^{140}Ba
Group 3	^{14}C, ^{18}F, ^{24}Na, ^{38}Cl, ^{32}P, ^{35}S, ^{42}K, ^{47}Ca ^{47}Sc, ^{51}Cr, ^{56}Mn, ^{59}Fe, ^{58}Co, ^{65}Zn, ^{69m}Zn ^{76}As, ^{77}As, ^{75}Se, ^{82}Br, ^{85m}Kr, ^{86}Rb, ^{85}Sr, ^{90}Y, ^{99}Mo, ^{99}Tc, ^{111}Ag, ^{109}Cd, ^{115}Cd, ^{115m}In, ^{113}Sn, ^{131m}Te, ^{132}Te, ^{132}I ^{131}Cs, ^{131}Ba, ^{140}La, ^{177}Lu, ^{186}Re, ^{188}Re, ^{198}Au, ^{199}Au, ^{197}Hg, ^{197m}Hg, ^{203}Hg
Group 4	^{3}H, ^{58m}Co, ^{69}Zn, ^{85}Kr, ^{85m}Sr, ^{91m}Y, ^{99m}Tc, ^{113m}In, ^{129}I, ^{131m}Xe, ^{133}Xe, ^{134m}Cs, ^{193m}Pt, ^{197m}Pt

toxic radionuclides. Since alpha and beta radiations are absorbed locally, intense radiation dose is delivered to small areas of tissue when alpha and beta emitters, particularly of long half-life are absorbed and retained in the body. Hence most long-lived alpha and beta emitters are categorized among the most toxic radionuclides. Radionuclides of elements which are not selectively retained in the body have low toxicity and are preferred for tracer applications, particularly for in vivo applications. Further, the shorter the half-life and the lower the energy of disintegration, the lower is the toxicity. Table 4 shows a classification of radionuclides, according to their relative radiotoxicities per unit activity, into four groups, namely Group 1: high toxicity, Group 2: medium-high toxicity, Group 3: medium toxicity, and Group 4: low toxicity. The specific activity of the tracer and its chemical form profoundly influence the toxicity; the lower the specific activity the lesser will be the radiation dose delivered per unit mass of absorbed material. Similarly, the shorter the biological half-life of the species, the lesser will be the dose delivered.

C. Purity
For tracer applications in the biomedical field three aspects of purity of the radionuclide are important, namely, radionuclidic purity, radiochemical purity, and chemical purity.

1. Radionuclidic Purity
Radionuclidic purity may be defined as the percentage of the total radioactivity in the preparation which is due to the stated radionuclide, daughter radionuclides being excluded. Usually, high radionuclidic purity, exceeding 99.9% is a prime requirement for the tracer since (1) even small levels of certain impurities might interfere in counting and quantitation of data and (2) for in vivo tracer applications, even small traces of impurities, particularly long-lived contaminants, might contribute to excessive radiation exposure to the patient. One study has shown,[1] for example, that contamination of $^{99}\Sigma$Tc with ^{134}Cs to the extent of 10^{-3}% would contribute an additional radiation burden of 50 to 85% of the ^{99m}Tc dose to the gonads and the liver. As a tentative guideline, it has been accepted that the radiological risk from impurities should be less than 10% of that from the principal radionuclide. In view of this, the requirement of radionuclidic purity for some of the short-lived radiotracers is extremely stringent.[1,2] Table 5 outlines the currently accepted specifications of radionuclidic purity for some reactor-produced tracer nuclides.[3-7]

Table 5
REQUIREMENTS OF RADIONUCLIDIC PURITY FOR REACTOR-PRODUCED RADIONUCLIDES

Radionuclide	Radionuclidic impurity	Main source of impurity	Acceptable upper limit of impurity
^{24}Na	^{42}K	(n, γ) reaction on K present in target	<1%
^{42}K	^{24}Na	(n, γ) reaction on Na present in target K_2CO_3 or $KHCO_3$	<0.5%
	^{36}Cl, ^{38}Cl, ^{35}S, ^{32}P	(n, γ), (n, p) and (n, α) reactions on chloride in KCl target	<0.01% each
	^{82}Br	(n, γ) reaction on Br present in target K_2CO_3 or $KHCO_3$	<0.1%
^{35}S	^{32}P	(n, α) reaction on Cl in target	<0.02%
	^{42}K, ^{36}Cl, ^{38}Cl	(n, γ) reactions in KCl target	<0.01% each
^{18}F	^{3}H	^{3}H from ^{6}Li (n, α) ^{3}H reaction	<2%
^{32}P	^{35}S	(n, γ) reaction on S	<0.1%
	^{33}P	(n, p) reaction on ^{33}S	<2%
^{45}Ca	^{47}Ca, ^{49}Ca	(n, γ) on ^{46}Ca, ^{48}Ca	<0.1% each
	^{47}Sc, ^{49}Sc	Daughter products of ^{47}Ca, ^{49}Ca	<0.01% each
^{58}Co	^{60}Co	(n, γ) on cobalt impurity in Ni target. Also ^{60}Ni (n, p) ^{60}Co reaction	<0.1%
^{86}Rb	^{134}Cs	(n, γ) on Cs impurity in target	<0.1%
	^{32}P, ^{35}S	(n, α) and (n, p) reactions on Cl in target RbCl	<0.1% each
^{59}Fe	^{55}Fe	(n, γ) on ^{54}Fe	<0.5%
	^{60}Co	(n, γ) on Co impurity in target	<0.001%
^{125}I	^{126}I	(n, γ) on ^{125}I	<1%
	^{137}Cs	^{136}Xe (n, γ)^{137}Xe $\xrightarrow{\beta^-}$ ^{137}Cs	<0.005%
^{99}Mo (n, γ) produced	^{60}Co, ^{65}Zn, ^{95}Zr, ^{124}Sb, ^{95}Nb, ^{192}Ir	(n, γ) reactions on impurities in the target MoO_3	$<2.5 \times 10^{-3}$% each
	^{110m}Ag	(n, γ) reaction on Ag impurity in target	$<5 \times 10^{-5}$%
	^{134}Cs	(n, γ) reaction on Cs impurity in target	$<10^{-5}$%
^{99}Mo (Fission produced)	^{131}I, ^{103}Ru	Fission products	$<2.5 \times 10^{-3}$%
	^{89}Sr	Fission product	$<3 \times 10^{-5}$%
	^{90}Sr	Fission product	$<3 \times 10^{-6}$%
	All other beta and gamma emitting impurities	Fission produced	$<5 \times 10^{-3}$%
	Gross alpha activity	(n, γ) reactions on ^{238}U etc.	$<5 \times 10^{-8}$%
^{99m}Tc (Prepared from neutron irradiated Mo)	^{99}Mo	Incomplete separation or bleeding of generator	<0.1%
	Other gamma emitting impurities	From impurities in ^{99}Mo	<0.05%

Table 5 (continued)
REQUIREMENTS OF RADIONUCLIDIC PURITY FOR REACTOR-PRODUCED RADIONUCLIDES

Radionuclide	Radionuclidic impurity	Main source of impurity	Acceptable upper limit of impurity
^{99m}Tc (Prepared from fission product ^{99}Mo)	^{99}Mo	Incomplete separation or bleeding of generator	<0.1%
	^{131}I	From impurities in ^{99}Mo	<5 × 10^{-3}%
	^{103}Ru	From impurities in ^{99}Mo	<5 × 10^{-3}%
	^{89}Sr	From impurities in ^{99}Mo	<6 × 10^{-5}%
	^{90}Sr	From impurities in ^{99}Mo	<6 × 10^{-6}%
	All other radionuclidic impurities (beta and gamma)	From impurities in ^{99}Mo	<10^{-2}%
	Gross alpha activity		<10^{-7}%

Table 6
REQUIREMENTS OF RADIOCHEMICAL PURITY FOR REACTOR-PRODUCED RADIONUCLIDES

Radionuclide	Chemical form	Impurity	Acceptable level of impurity
^{18}F	Fluoride in dilute alkali	AlF_6^{3-}	<1%
^{32}P	Orthophosphate in dilute HCl	Meta, Pyro and Poly phosphate	Together <1%
^{51}Cr	Chromate in NaCl solution	Cr^{3+}, colloidal $Cr(OH)_3$	Together <2%
^{51}Cr	Chromic chloride	$CrO_4^=$ colloidal $Cr(OH)_3$	<1% <0.1%
^{99m}Tc	Pertechnetate in NaCl solution	Reduced ^{99m}Tc species (cationic)	Together <2%
^{125}I, ^{131}I	Iodide in dilute alkali	IO_3^-, IO_4^- organically bound I	Together <1% <0.1%

2. Radiochemical Purity

High radiochemical purity is an important requirement for nuclides used as radiotracers or as intermediates for the preparation of labeled compounds. Radiochemical purity may be defined as the percentage of the total radioactivity that is present in the stated chemical form. For example, in a preparation of ^{51}Cr in the form of sodium chromate, radiochemical purity would imply that not less than 95% of the ^{51}Cr activity should be present in the chemical form of chromate, and all other chemical forms, such as $^{3+}$Cr, colloidal chromic hydroxide, etc., should not totally exceed 5% of the radioactivity.[8-10] The above specification is important since only the chromate form can tag on to red blood cells while trivalent chromium is not available for RBC labeling. Similarly high radiochemical purity is an important requirement for ^{131}I employed as a tracer for thyroid studies or as a starting material for the preparation of labeled compounds; the ^{131}I should be present in the form of iodide as other chemical forms such as iodate and periodate do not afford high labeling yields.[11] Table 6 outlines the currently accepted specifications of radiochemical purity for some reactor-produced tracer nuclides.

Table 7
REQUIREMENTS OF CHEMICAL PURITY FOR SOME REACTOR-PRODUCED RADIONUCLIDES

Radionuclide and chemical form	Chemical impurity	Acceptable limit of impurity
^{18}F	Al	<2 µg/mℓ
Fluoride in dil. alkali	Heavy metals	<5 µg/mℓ
^{35}S	Heavy metals	<5 µg/mℓ
Sulfate in dil. HCl	Total solids	<1 mg/mℓ
^{32}P	Heavy metals	<5 µg/mℓ
O-phosphoric acid in dil. HCl	Sulfate	<100 µg/mℓ
	Total solids	<1 mg/mℓ
	Hydroxide precipitable matter (pH 6.7—8)	Negligible
^{51}Cr	As, Pb, Al	<5 µg/mℓ each
Chromate in NaCl solution	Total solids	<10 mg/mCi ^{51}Cr
	Metallic impurities (other than Cr)	<10 µg/mℓ
^{99m}Tc	Al	<2 µg/mℓ (15 mCi)
Pertechnetate in NaCl solution	Mo	<50 µg/mℓ
	Heavy metals	<5 µg/mℓ
^{125}I	Heavy metals	<5 µg/mℓ
Iodide in dil. NaOH	Total solids	<5 mg/mℓ
	Reducing agents	$<5 \times 10^{-4}$ N in sulfite
^{131}I	As, Pb, Se	<5 µg/mℓ
Iodide in dil. NaOH solution	Te	<2 µg/mℓ
	Other metals	<1 µg/mℓ
	Reducing agents	$<5 \times 10^{-4}$ N in sulfite

3. Chemical Purity

High chemical purity is an important requirement for radionuclides employed as tracers or as intermediates for the preparation of labeled compounds. Chemical purity may be defined as the percentage of the specified chemical form in the preparation regardless of any isotopic substitution, vehicles, carriers, preservatives, and essential additives being excluded. However, the chemical concentration levels of most radiotracer preparations are very small and are in the range of micro or millimolar and are often below the concentration levels of many other ingredients which may be present. For example, in a solution of sodium iodide ^{131}I of concentration 10 mCi/mℓ, the concentration of NaI will be in the range of 2 to 10 µM whereas that of carbonate, chloride, sulfate, etc., may be in the range of 10 to 100 µM. The chemical purity of such preparations is hence specified in terms of the absence of undesirable impurities such as heavy metal contaminants (these should be below 2 parts per million of the solution), and reducing agents such as oxalate, thiosulfate, sulfite, etc. The latter are undesirable in solutions of ^{125}I and ^{131}I used for labeling of proteins and hormones, but do not interfere in the use of these nuclides for thyroid function studies. Table 7 outlines the currently accepted criteria of chemical purity for some reactor-produced radionuclides.

D. High Specific Activity

High specific activity is an important requirement for the radiotracer for most applications in biology and medicine. It has been found that ^{131}I in the form of NaI, employed as a tracer for thyroid investigations should have a minimum specific activity of 3 mCi/µg I, as the presence of higher levels of carrier iodine tends to reduce the uptake of ^{131}I by the thyroid gland.[12] High specific activity is also required for isotopes meant to be used for labeling

Table 8
SPECIFIC ACTIVITY OF SOME REACTOR-PRODUCED RADIONUCLIDES

Radionuclide	Maximum theoretical specific activity Ci/g	Practicable maximum specific activity Ci/g
^{3}H	9.71×10^3	9×10^3
^{14}C	4.46	3.8
^{24}Na	8.7×10^6	1
^{32}P	2.8×10^5	2.5×10^4
^{35}S	4.3×10^4	10^4
^{42}K	5.9×10^5	0.5
^{45}Ca	1.78×10^4	0.5
^{51}Cr	9.27×10^4	1500
^{58}Co	3.13×10^4	>3000
^{59}Fe	4.8×10^4	50
^{99}Mo	4.8×10^5	$>5 \times 10^4$
^{86}Rb	8.2×10^4	0.3
^{125}I	1.72×10^4	1.5×10^4
^{131}I	1.25×10^5	2×10^4

hormones and other biological substances for radioimmunoassays and other related assay procedures.[11,13-15] For example, it has been found that optimum yields and specific activities of ^{125}I labeled hormones such as triiodo-L-thyroxine and L-thyroxine to be used in radioimmunoassays are obtained only if the ^{125}I employed for labeling is of very high specific activity (exceeding 15 curies per milligram and nuclidic abundance exceeding 90%). However, often problems are encountered in solutions of high specific activity radionuclides due to (1) instability of the chemical form in which the radionuclide is present or (2) loss of the tracer due to adsorption, or due to chemical or microbiological reactions. For example, in solutions containing ^{131}I of high specific activity (exceeding 25 curies per milligram I) in the form of iodide, the iodine is readily oxidized to iodate and periodate by self-radiolysis.[16] The rate of oxidation depends on the pH of the solution, the specific activity of the ^{131}I and the radioactive concentration. Another example is instability of high specific activity chromate-^{51}Cr which tends to get reduced in aqueous solution due to interaction with trace impurities.[17-32] In neutral and alkaline solutions of high specific activity in the chemical form of phosphate, adsorption of ^{32}P onto the walls of the glass container is often a serious problem. Preparations of high specific activity Ba^{14}CO$_3$ and Na$_2$^{14}CO$_3$ can undergo exchange reactions with atmospheric CO$_2$.[17] Microbiological attack can cause decomposition of solutions of sodium phosphate-^{32}P, and the effect is pronounced if the specific activity is very high. In many cases, the addition of certain preservatives, such as sodium thiosulfate in solutions of ^{131}I, or hold-back carrier will minimize these effects.[16] Table 8 gives the theoretical maximum specific activities of a few important reactor-produced radionuclides and the practicable specific activities for tracer applications.

E. Ready Availability and Low Price

Ready availability and low price are important considerations in the choice of radionuclides for tracer applications. The availability and price of reactor-produced radionuclides depends on a variety of factors, namely (1) availability and cost of the "targets" required for reactor irradiation and of the chemicals and materials required for processing and purification, (2) the rate of production of the radionuclide, (3) the nature and quantity of radioactive wastes produced, and (4) the pattern of demand for the radionuclide. The target materials for reactor irradiation and the chemicals and equipment required for processing and purification of the radionuclide should be abundantly available and inexpensive. The cross-section for the nuclear reaction giving rise to the desired radionuclide should be sufficiently high so that large quantities can be produced in a single irradiation container. The production of large quantities of radioactive wastes originating from concurrent nuclear reactions, would make the production process more difficult and expensive as these wastes have to be properly stored and safely disposed of. As an example, in the production of ^{99}Mo by the direct neutron irradiation of pure MoO_3, no long-lived waste products are generated and hence the cost of production and purification is low. However, production and purification of ^{99}Mo from the fission of ^{235}U would involve the handling of substantial quantities of long-lived isotopes such as ^{134}Cs, ^{103}Ru, ^{90}Sr, etc., hence, this process is more expensive. Moreover, the requirements of quality control for fission produced ^{99}Mo are more stringent than for (n,γ)produced ^{99}Mo and hence the cost of production is likely to be correspondingly higher. The pattern of demand has an important bearing on the economics of production. Radionuclides which have assured regular demands can be processed and kept in stock, thus ensuring their ready availability. Products which are required only occasionally and in small quantities may have to be "made to order". Such processing is expensive. Table 9 shows the current availability and price of some of the important reactor-produced radionuclides.

F. Special Requirements for Preparation of Labeled Compounds and Radiopharmaceuticals

Many reactor-produced radionuclides are used for preparation of labeled compounds which are in turn employed as tracers in biology and medicine. Some of the radionuclides are also employed for the preparation of radiopharmaceutical formulations which are used in nuclear medicine for in vivo or in vitro studies. ^{3}H, ^{14}C, ^{35}S, ^{32}P, ^{51}Cr, ^{75}Se, ^{99m}Tc, ^{125}I, and ^{131}I are the main reactor-produced radionuclides which are employed for the preparation of labeled compounds and radiopharmaceuticals. For such applications, some of the radionuclides employed should, in addition to meeting the criteria listed in Section II, fulfill certain additional requirements. For example, ^{125}I and ^{131}I to be used for "protein iodination" should be free from reducing agents such as sulfite, thiosulfate, oxalate, etc. Even traces of unknown impurities present in preparations of these nuclides[14,15,18] and H_2O_2 produced by self-radiolysis interfere in the labeling and denature the labeled proteins. The biological and immunological properties of many proteins, hormones, and antigens are adversely affected by trace impurities present in the labeling radionuclide. Successful labeling of these products for applications in metabolic studies or in radioimmunoassays would require the use of ^{125}I and ^{131}I of very high purity and specific activity. It has been found that freshly prepared aqueous or dilute NaOH solutions of ^{131}I and ^{125}I of high radioactive concentration (exceeding 300 mCi/mℓ) and specific activity (exceeding 15 Ci/mg) give good labeling yields without loss of biological and immunological properties. It may be mentioned that sometimes traces of impurities may be leached out from containers, and serious adsorption losses may be encountered when solutions of such high radioactive concentrations and specific activities are handled. Aging of such preparations usually results in lower yields in labeling and labeled products of unsatisfactory purity.

^{99}Mo and ^{113}Sn are employed for the preparation of generators for ^{99m}Tc and ^{113m}In, respectively. For such use, these isotopes are required with very high specific activity and radioactive concentration, since bleeding of the parent isotopes is likely to take place during elution of ^{99m}Tc and ^{113m}In if the adsorption of the parent is not confined to a narrow band on the top of the column. For Mo^{99} the requirements are specific activity greater than 10 Ci/g, and radioactive concentration greater than 100 mCi/mℓ. However, material of specific activity greater than 50 Ci/mg is now available from the fission of ^{235}U and is preferred for preparation of column generators. For ^{113}Sn, the optimum specific activity required is 15 Ci/g. Further it has been found that some of the chemicals and reagents added in the processing of the parent isotopes are carried over with the daughter nuclides and give rise to problems in the preparation of their labeled compounds. For example, H_2O_2 added for the dissolution

Table 9
CURRENT AVAILABILITY AND TYPICAL PRICES OF SOME IMPORTANT REACTOR-PRODUCED RADIONUCLIDES

Radionuclide	Availability	Price in U.S. $
^{3}H	Readily available as ^{3}H gas, tritiated water	Up to several curies $15 per Ci. Large quantities $9/Ci
^{14}C	Readily available in the form of $BaCO_3$ and CO_2	Up to a few millicuries $20/m Ci. Large quantities $15/m Ci
^{18}F	Available only from a few suppliers	
^{24}Na	Readily available as NaCl solution and as irradiated Na_2CO_3	$25/mCi
^{42}K	Readily available as KCl solution and as irradiated K_2CO_3 or KCl	$200/mCi (For high specific activity material)
^{32}P	Readily available as O — Phosphoric acid in dilute HCl	Up to a few millicuries $12/m Ci. Larger quantities $3.50/m Ci
^{35}S	Readily available as sulfuric acid in dilute HCl or as sulfate in aqueous solution	$14/m Ci
^{51}Cr	Readily available as $CrCl_3$ in dilute HCl or as sodium chromate in aqueous or NaCl solution	$15/mCi for highest specific activity material
^{58}Co	Readily available as cobaltous chloride in dilute HCl	$24/mCi.
^{59}Fe	Readily available as ferric chloride in dilute HCl	$100/mCi
^{45}Ca	Readily available as $CaCl_2$ in dilute HCl	$30/mCi.
^{36}Cl	Readily available as dilute HCl solution	$160 for 100 μCi
^{82}Br	Readily available as NH_4Br or NaBr solution or as irradiated KBr	$50/mCi
^{99}Mo	Readily available as	
	(i) Ammonium molybdate in dilute NH_4OH solution (produced by (n, γ) activation of MoO_3)	$90/100 mCi
	(ii) Generator containing ^{99}Mo (produced by (n, γ) activation of MoO_3)	$120/100 mCi
	(iii) Ammonium molybdate in dilute NH_4OH solution (fission produced ^{99}Mo)	$100/100 mCi
	(iv) Generator for ^{99m}Tc (fission produced ^{99}Mo)	$150/100 mCi
^{113}Sn	Readily available as $SnCl_2$ in dilute HCl or as generator for ^{113m}In	$1500/50 mCi
^{113m}In	Readily available as a generator containing the parent ^{113}Sn	$1500/50 mCi

Table 9 (continued)
CURRENT AVAILABILITY AND TYPICAL PRICES OF SOME IMPORTANT REACTOR-PRODUCED RADIONUCLIDES

Radio-nuclide	Availability	Price in U.S. $
^{125}I	Readily available as iodide in dilute NaOH solution with high radioactive concentration (free from reducing agents)	$12/mCi (for large quantities)
^{131}I	Readily available as (i) iodide in dilute NaOH solution with hgh radioactive concentration free from reducing agents and (ii) as sodium iodide in aqueous solution containing sodium thiosulfate	$7/mCi (for large quantities)

of irradiated MoO_3 is sometimes eluted along with the pertechnetate and interferes with the reduction of TcO_4^- during the preparation of labeled formulations.

Radionuclides used for the preparation of in vivo injectable radiopharmaceuticals should be stored under conditions which will prevent the growth of microorganisms. This is important for ^{131}I and ^{32}P.

III. MAIN PRODUCTION REACTIONS

Six main nuclear reactions induced by neutrons are employed for regular and large-scale production of radionuclides of biomedical interest in nuclear reactors. These are (1) (n,γ) reaction, (2) (n,γ) followed by decay (3) (n,p) reaction, (4) (n,α) reaction, (5) (n, fission) reaction, and (6) successive neutron capture reactions, followed by a decay in some cases. These can be represented as X(n,γ) Y(n,γ) Z decay A. In addition to the above, the Szilard-Chalmers recoil enrichment process following $\overrightarrow{(n,\gamma)}$ reactions in some stable chemical compounds and indirect reactions such as (t, p) and (t, n) caused by energetic tritons produced by the neutron-induced fission of lithium-6 are also of some interest for production of a few useful radionuclides of high specific activity. These reactions are described in some detail below.

A. (n,γ) Reactions

These reactions are generally caused by thermal neutrons, that is, neutrons of energy in the range of 0.025 eV. A large number of elements interact with thermal neutrons by the (n,γ) reaction and useful radionuclides are produced.

Typical examples of (n,γ) process are

$$^{23}_{11}Na + ^{1}_{0}n \longrightarrow \begin{cases} ^{24m}_{11}Na + \gamma \\ ^{24}_{11}Na + \gamma \end{cases}$$

$$^{98}_{42}Mo + ^{1}_{0}n \longrightarrow ^{99}_{42}Mo + \gamma$$

$$^{58}_{26}Fe + ^{1}_{0}n \longrightarrow ^{59}_{26}Fe + \gamma$$

In the first reaction ^{23}Na which forms 100% of natural sodium, captures a thermal neutron and produces the 20.21m half-life ^{24m}Na and the 15.03 hr half-life ^{24}Na. The activation cross-sections for these reactions are 0.43 and 0.10 barns, respectively. ^{24m}Na decays by isomeric transition to ^{24}Na. No other radionuclide of sodium is produced in this process. In the third reaction ^{58}Fe which forms 0.29% of natural iron undergoes (n,γ) activation to produce the 44.56 d half-life ^{59}Fe. A concurrent (n,γ) reaction on ^{54}Fe (natural abundance 5.8%) namely ^{54}Fe (n,γ) ^{55}Fe (half-life 2.68 years) also takes place unless pure ^{58}Fe is used as the target. Another reaction, namely ^{54}Fe (n, p) ^{54}Mn (312.2 d) can take place with neutrons of higher energy (exceeding 2 MeV). The cross-sections for these reactions are as follows:

^{58}Fe (n, γ) ^{59}Fe	1.14 barns
^{54}Fe (n, γ) ^{55}Fe	2.2 barns
^{54}Fe (n, p) ^{54}Mn	300 millibarns (4 MeV)

Hence, when natural iron is irradiated in a nuclear reactor, a mixture of ^{59}Fe, ^{55}Fe, and ^{54}Mn will be produced. A rigorous calculation using Equation 1 given under Section IV would, however, show that ^{54}Mn will be present in negligibly small amounts in the mixture.

The radionuclide produced by the (n, γ) reaction is isotopic with the capturing nucleus; hence it is not usually possible to separate the radionuclide thus produced from the stable isotope or isotopes constituting the target element. In a few cases, however, when stable chemical compounds of certain elements are irradiated, a Szilard-Chalmers recoil process takes place and a significant percentage of the radionuclide produced by the (n, γ) reaction appears in a chemical form different from that of the target. In these cases, the fraction of the radionuclide which appears in the "recoil chemical form" can be separated with high specific activity (see Section III.G.)

The activation cross-sections for many (n, γ) reactions are high and are in the range of several barns. Table 10 gives the (n, γ) activation cross-sections for some target nuclides leading to production of useful radionuclides.

Generally, the total activity of the radionuclide produced in the target will be directly proportional to the mass of the target, the neutron flux, and the activation cross-section; the total activity will increase with time of irradiation and will attain a saturation value depending on the half-life of the radionuclide. In stating the above, it is assumed that (1) the target material does not have a very high neutron absorption cross-section, (2) the product radionuclide does not have a high neutron absorption cross-section, and (3) the period of irradiation is short compared to the half-life of the product radionuclide. A more detailed treatment of the various factors involved in calculation of the activation yield is given in Section IV. Table 11 lists the important reactor-produced radionuclides, produced by (n, γ) reactions.

Though (n, γ) reactions are the simplest and most abundant among the production processes in a reactor, they have certain limitations. Since in (n, γ) reactions, excluding cases where Szilard-Chalmers enrichment takes place, the radionuclide produced cannot be separated from the stable isotopes of the capturing element, the specific activity attained will be low compared to the theoretical specific activity of the "carrier-free" radionuclide (Table 8). Enriched targets can be used in some cases to obtain high specific activities. In many targets concurrent (n, γ) reactions may take place giving rise to radionuclidic impurities. In certain other cases, the radionuclide produced may undergo a further (n, γ) reaction leading to the production of an impurity. A few such instances where radionuclidic impurities are produced in (n, γ) reactions are shown in Table 12.

B. (n, γ) Followed by Decay

In this reaction, the host nucleus captures a thermal neutron giving rise to a radionuclide intermediate which decays to produce the radionuclide of interest. Important examples of such processes are

Table 10
(n, γ) ACTIVATION CROSS-SECTIONS FOR SOME TARGET NUCLIDES

Target nuclide	Activation cross-section in barns	Radionuclide produced
^{23}NA	0.10	^{23}Na
^{23}Na	0.43	^{24m}Na
^{41}K	1.46	^{42}K
^{44}Ca	0.88	^{45}Ca
^{46}Ca	0.70	^{47}Ca
^{50}Cr	15.9	^{51}Cr
^{81}Br	2.7	^{82}Br
^{84}Sr	0.3	^{85}Sr
^{84}Sr	0.59	^{85m}Sr
^{85}Rb	0.40	^{86}Rb
^{85}Rb	0.047	^{86m}Rb
^{34}S	0.24	^{35}S
^{74}Se	52	^{75}Se
^{58}Fe	1.14	^{59}Fe
^{54}Fe	2.20	^{55}Fe
^{37}Cl	0.428	^{38}Cl
^{37}Cl	0.005	^{38m}Cl
^{98}Mo	0.13	^{99}Mo
^{196}Hg	3×10^3	^{197}Hg
^{196}Hg	120	^{197m}Hg
^{202}Hg	5.0	^{203}Hg
^{124}Xe	100	^{125}Xe
^{124}Xe	20	^{125m}Xe
^{130}Te	0.2	^{131}Te
^{130}Te	0.03	^{131m}Te
^{125}I	900	^{126}I
^{168}Yb	3.5×10^3	^{169}Yb
^{185}Re	110	^{186}Re
^{185}Re	0.3	^{186m}Re
^{187}Re	74	^{188}Re
^{187}Re	1.0	^{188m}Re
^{112}Sn	0.306	^{113m}Sn
^{112}Sn	0.40	^{113}Sn

$$^{130}_{52}\text{Te} + ^{1}_{0}\text{n} \longrightarrow \begin{array}{c} ^{131}_{52}\text{Te} \\ \uparrow \text{IT} \\ ^{131m}_{52}\text{Te} \end{array} \xrightarrow{\beta^-} {}^{131}_{53}\text{I}$$

$$^{124}_{54}\text{Xe} + ^{1}_{0}\text{n} \longrightarrow \begin{array}{c} ^{125}_{54}\text{Xe} \\ \uparrow \text{IT} \\ ^{125m}_{54}\text{Xe} \end{array} \xrightarrow{EC, \beta^+} {}^{125}_{53}\text{I}$$

Since the radionuclide produced by the (n, γ) followed by decay process can be readily separated from the target, specific activities approaching the "carrier-free" level can be attained by this process. For attaining the highest specific activities for the end product, the following factors have to be ensured. (1) The target material should be free from stable nuclides of the product radionuclide. For example, in the production of ^{131}I by irradiation of tellurium targets, the target should not be contaminated with stable iodine. As a corollary to this, the chemicals and reagents employed in the processing of the irradiated target should

Table 11
REACTOR-PRODUCED RADIONUCLIDES PRODUCED BY (n, γ) REACTIONS

Radionuclide	Target/Targets used for (n, γ) activation	Remarks
^{24}Na	Na_2CO_3, $NaHCO_3$	
^{42}K	K_2CO_3, KCl	KCl target can be irradiated at high fluxes
^{45}Ca	$CaCO_3$ (natural) $^{44}CaCO_3$ (enriched)	
^{51}Cr	$^{50}Cr_2O_3$, (metal) K_2CrO_4	High flux irradiation required Szilard-Chalmers enrichment
^{59}Fe	$^{58}Fe_2O_3$, Fe^{58} (metal)	^{55}Fe impurity produced will depend on the content of ^{54}Fe in $^{58}Fe_2O_3$
^{64}Cu	CuO Cu-phthalocyanine	Szilard-Chalmers enrichment possible
^{82}Br	NH_4Br	Only low flux irradiation possible due to instability of NH_4Br
	KBr	High flux irradiation possible. ^{42}K impurity is produced
^{86}Rb	Rb_2CO_3, RbCl	RbCl can be irradiated at high fluxes
^{99}Mo	MoO_3, $^{98}MoO_3$	The maximum specific activity attainable is low, <15 Ci/g
^{113}Sn	Sn metal or SnO_2 enriched in ^{112}Sn	

also be free from stable iodine contamination. (2) No stable or long-lived nuclides of the product radionuclide should be produced by concurrent nuclear reactions. If these are produced, the specific activity of the product radionuclide of interest will be diluted. For example, in the production of ^{131}I by (n, γ) decay reaction on tellurium, ^{127}I, ^{129}I are also produced by the following concurrent reactions:

$$^{126}_{52}Te + ^{1}_{0}n \longrightarrow \begin{array}{c} ^{127}_{52}Te \\ \uparrow IT \\ ^{127m}_{52}Te \end{array} \xrightarrow{\beta^-} ^{127}_{53}I \text{ (stable)}$$

$$^{128}_{52}Te + ^{1}_{0}n \longrightarrow \begin{array}{c} ^{129}_{52}Te \\ \uparrow IT \\ ^{129m}_{52}Te \end{array} \xrightarrow{\beta^-} ^{129}_{53}I \text{ (half-life } 1.57 \times 10^7 \text{ years)}$$

The activities and masses of these isotopes produced under 2 sets of irradiation conditions are given in Table 13. The use of enriched tellurium targets will minimize or eliminate the production of such carrier isotopes which would dilute the specific activity of the product radionuclide. Table 14 lists the important radionuclides which are obtained by (n, γ) decay processes.

C. (n, p) and (n, α) Reactions

Here the host nuclide incorporates a neutron (generally a fast neutron) and a proton or alpha-particle is ejected, giving rise to the desired radionuclide. This radionuclide being

Table 12
PRODUCTION OF RADIONUCLIDIC IMPURITIES IN (n, γ) REACTIONS

Reaction of interest and required radionuclide	Side reactions producing radionuclidic impurities	Radionuclidic impurities produced
^{23}Na (n, γ) ^{24}Na; ^{24}Na	^{41}K (n, γ) ^{42}K (K impurity in target)	^{42}K
^{41}K (n, γ) ^{42}K; ^{42}K	^{23}Na (n, γ) ^{24}Na (Na impurity in target)	^{24}Na
	^{81}Br (n, γ) ^{82}Br	^{80}Br, ^{80m}Br, ^{82}Br
	^{79}Br (n, γ) ^{80}Br, ^{80m}Br (Br impurity in target)	
	^{35}Cl (n, γ) ^{36}Cl	^{36}Cl, ^{38}Cl,
	^{37}Cl (n, γ) ^{38}Cl	^{35}S, ^{32}P
	^{35}Cl (n, p) ^{35}S	
	^{35}Cl (n, α) ^{32}P	
	(From Cl in KCl target)	
^{44}Ca, (n, γ) ^{45}Ca; ^{45}Ca	^{46}Ca (n, γ) ^{47}Ca	^{47}Ca, ^{49}Ca ^{47}Sc, ^{49}Sc
	^{47}Ca $\xrightarrow{\beta^-}$ ^{47}Sc	
	^{48}Ca (n, γ) ^{49}Ca	
	^{49}Ca $\xrightarrow{\beta^-}$ ^{49}Sc	
^{85}Rb (n, γ) ^{86}Rb; ^{86}Rb	^{133}Cs (n, γ) ^{134}Cs (from Cs impurity in target)	^{134}Cs
	^{35}Cl (n, p) ^{35}S	^{35}S, ^{32}P
	^{35}Cl (n, α) ^{32}P	
	(from Cl in RbCl target)	
^{58}Fe(n,γ) ^{59}Fe	^{54}Fe(n,γ) ^{55}Fe	^{55}Fe
^{59}Fe	^{59}Co(n,γ) ^{60}Co (from Co-impurity in target)	^{60}Co
^{98}Mo(n,γ) ^{99}Mo	^{59}Co(n,γ) ^{60}Co	^{60}Co, ^{65}Zn, ^{95}Zr, ^{95}Nb,
^{99}Mo	^{64}Zn(n,γ) ^{65}Zn	^{10m}Ag, ^{134}Cs, ^{186}Re,
	^{94}Zr(n,γ) ^{95}Zr	^{188}Re, ^{192}Ir
	^{95}Zr $\xrightarrow{\beta^-}$ ^{95}Nb	
	^{109}Ag(n,γ)^{110m}Ag	
	^{133}Cs(n,γ) ^{134}Cs	
	^{185}Re(n,γ) ^{186}Re	
	^{187}Re(n,γ) ^{188}Re	
	^{191}Ir(n,γ) ^{192}Ir	
	(from the impurities in the target)	
^{112}Sn(n,γ) ^{113}Sn	^{124}Sn(n,γ)^{125}Sn $\xrightarrow{\beta^-}$ ^{125}Sb	^{125}Sn, ^{125}Sb, ^{124}Sb, ^{122}Sb,
^{113}Sn		^{60}Co, ^{51}Cr, ^{59}Fe, ^{55}Fe,
	^{121}Sb(n,γ) ^{122}Sb	^{65}Zn, ^{192}Ir
	^{123}Sb(n,γ) ^{124}Sb	
	^{59}Co(n,γ) ^{60}Co	
	^{50}Cr(n,γ) ^{51}Cr	
	^{58}Fe(n,γ) ^{59}Fe	
	^{54}Fe(n,γ) ^{55}Fe	
	^{64}Zn(n,γ) ^{65}Zn	
^{191}Ir	^{191}Ir(n,γ) ^{192}Ir (from impurities in target)	
^{197}Au(n,γ) ^{198}Au	^{198}Au(n,γ) ^{199}Au	^{199}Au
^{198}Au		

Table 13
MASSES OF IODINE ISOTOPES AND ACTIVITIES OF ^{131}I PRODUCED IN NEUTRON IRRADIATION OF NATURAL TELLURIUM

Time of irradiation	Percentage by weight of I isotopes produced			Total weight in microgram of I per 100 mCi ^{131}I	Sp. activity of ^{131}I Ci/mg
	^{127}I	^{129}I	^{131}I		
10 days	61	20	19	4.2	23.8
20 days	66	20	14	5.9	16.95

Table 14
IMPORTANT RADIONUCLIDES PRODUCED BY (n, γ) FOLLOWED BY DECAY

Radionuclide of interest	Nuclear reaction/reactions
^{99m}Tc	^{98}Mo (n, γ) ^{99}Mo $\xrightarrow{\beta^-}$ ^{99m}Tc
^{111}Ag	^{110}Pd (n, γ) ^{111}Pd, ^{111m}Pd $\xrightarrow{\beta^-}$ ^{111}Ag
^{113m}In	^{112}Sn (n, γ) ^{113}Sn $\xrightarrow{EC}$ ^{113m}In
^{115m}In	^{114}Cd (n, γ) ^{115}Cd $\xrightarrow{\beta^-}$ ^{115m}In
^{125}Sb (parent of ^{125m}Te)	^{124}Sn (n, γ) ^{125}Sn $\xrightarrow{\beta^-}$ ^{125}Sb
^{125}I	^{124}Xe (n, γ) ^{125}Xe $\xrightarrow{EC}$ ^{125}I
^{131}I	^{130}Te, (n, γ) ^{131}Te, ^{131m}Te $\xrightarrow{\beta^-}$ ^{131}I
^{131}Cs	^{130}Ba (n, γ) ^{131}Ba $\xrightarrow{EC}$ ^{131}Cs
^{199}Au	^{198}Pt (n, γ) ^{199}Pt $\xrightarrow{\beta^-}$ ^{199}Au

chemically different from the target element can be readily separated from the target and obtained as a high specific activity preparation. Examples of such process are

$$^{32}_{16}S + ^{1}_{0}n \longrightarrow ^{32}_{15}P + ^{1}_{1}H \quad (n,p) \text{ reaction}$$

$$^{27}_{13}Al + ^{1}_{0}n \longrightarrow ^{24}_{11}Na + ^{4}_{2}He \quad (n,\alpha) \text{ reaction}$$

The cross-sections for such reactions are small and are usually of the order of millibarns to a few barns. As in the case of (n, γ) followed by decay, the highest specific activity can be attained for the product radionuclide only if the target is initially free from stable isotopes of the product radionuclide and these are not produced by concurrent nuclear reactions. Since only a relatively small fraction of the neutron flux in a reactor will have energy in excess of the threshold value required for (n, p) and (n, α) reactions and since the cross-sections for these reactions are low, the rate of production in these cases will be considerably lower than in (n, γ) reactions. Table 15 lists the important radionuclides of interest as tracers produced by (n, p) and (n, α) reactions.

D. Fission

Neutron capture fission of U-235 and Pu-239 gives rise to a wide variety of fission product radionuclides, many of which are of interest as biomedical tracers. Their atomic numbers range from 30 to 66. While the long-lived fission product nuclides such as Cs-137, Sr-90,

Table 15
(n, p) AND (n, α) REACTIONS FOR PRODUCTION OF RADIONUCLIDES FOR TRACER APPLICATIONS

Reaction and cross-section	Target	Radionuclide produced
$^6_3Li(n, \alpha)^3_1H$ 950 barns[1] (thermal neutrons)	Li_2CO_3	^3H
$^{14}_7N(n, p)^{14}_6C$ 1.81 barns (for thermal neutrons)	AlN, Be_3N_2	^{14}C
$^{35}_{17}Cl(n, p)^{35}_{16}S$ 107 millibarns (for 14.1 MeV neutrons)	KCl	^{35}S
$^{32}_{16}S(n,p)^{32}_{15}P$ 60 millibarns for fast neutrons (225 millibarns for 14.1 MeV neutrons)	S	^{32}P
$^{58}_{28}Ni(n,p)^{58}_{27}Co$ 410 millibarns (for 14.1 MeV neutrons)	Ni	^{58}Co

Table 16
IMPORTANT FISSION PRODUCED RADIONUCLIDES OF INTEREST AS BIOMEDICAL TRACERS

Radionuclide	Cumulative fission yield (^{235}U fission)
^{99}Mo	6.14%
^{90}Y	5.74%
(Daughter of ^{90}Sr)	5.89% for ^{90}Sr
^{91}Y	6.01%
^{137m}Ba	6.18% (^{137}Cs)
(Daughter of ^{137}Cs)	
^{131}I	2.835%
^{140}La	6.28% for ^{140}Ba
(Daughter of ^{140}Ba)	

etc., can be conveniently recovered from fuel reprocessing wastes, medium and short-lived radionuclides are usually produced by irradiation of uranium (natural or enriched) in the reactor.

Table 16 lists the important fission product radionuclides of interest as biomedical tracers; their fission yields are also given in the table.

It may be mentioned that since the fission process gives rise to a mixture of radionuclides some of which are highly toxic (e.g., ^{90}Sr, ^{137}Cs), stringent purificatory procedures have necessarily to be employed to recover the radionuclides of interest with the requisite purity. Rigorous quality controls have also to be instituted to confirm the absence of long-lived radionuclidic contaminants. These aspects are discussed in greater detail in Section VII.

E. Successive Neutron Capture Reactions

Successive capture of two or more neutrons, sometimes with beta decay processes also involved are of interest for the production of some useful radionuclides for tracer applications. An example is the production of high specific activity ^{188}W which is the parent of ^{188}Re. ^{188}Re has potential applications in medicine.

$$W^{186}_{74} + n^1_0 \xrightarrow{(37.8b)} W^{187}_{74} \xrightarrow[(64b)]{(n,\gamma)} W^{188}_{74} \xrightarrow{\beta^-} Re^{188}_{75}$$
$$(28.6\%) \qquad\qquad (23.8\ h) \qquad (69\ d) \qquad (16.98\ h)$$

Another possible reaction which has been studied is

$$S^{36}_{16} \xrightarrow{(n,\gamma)} S^{37}_{16} \xrightarrow{(n,\gamma)} S^{38}_{16} \xrightarrow{\beta^-} Cl^{38}_{17}$$
$$(0.015\%) \quad (5.1\ m) \quad (2.83\ h) \quad (37.18\ m)$$

^{38}S may be used as the parent for ^{38}Cl in a "nuclide generator".

In two other instances, however, successive neutron capture reactions are the source of radionuclidic impurities as in the case of ^{125}I and ^{198}Au. The impurities produced in these cases are ^{126}I and ^{199}Au, respectively.

F. Indirect Reactions

Energetic tritons, having energy 2.73 MeV, produced by the neutron-induced fission of ^{6}Li can interact with nuclei of lighter elements to produce useful radionuclides. The two practicable examples are

$$O^{16}_8 + H^3_1 \longrightarrow F^{18}_9 + n^1_0 \qquad (t,n)\ \text{reaction}$$
$$(99.75\%) \qquad\quad (109.7\ m)$$

$$Mg^{26}_{12} + H^3_1 \longrightarrow Mg^{28}_{12} + H^1_1 \qquad (t,p)\ \text{reaction}$$
$$(11.01\%) \qquad\quad (20.92h)$$

G. Szilard-Chalmers Recoil Enrichment

When stable compounds of some elements are irradiated with thermal neutrons, a significant fraction of the radionuclide produced by the (n, γ) reaction appears in a chemical form different from that of the target. This fraction, known as the recoiled chemical form, can be readily separated from the target by simple radiochemical techniques. For example, when pure crystalline K_2CrO_4 is irradiated with thermal neutrons, 10 to 50% of the ^{51}Cr produced by the ^{50}Cr (n, γ) ^{51}Cr reaction appears in the trivalent state. The trivalent ^{51}Cr activity can be readily separated from the K_2CrO_4 target by dissolution followed by either direct or carrier precipitation of Cr^{3+} as hydroxide[19] or by adsorption of Cr^{3+} on to Al_2O_3.[20] If the K_2CrO_4 target is initially free from Cr^{3+} impurity and if no significant amount of Cr^{3+} is produced by decomposition of K_2CrO_4 (resulting from interactions with gamma radiations or fast neutrons present in the reactor), the trivalent ^{51}Cr produced will be of high specific activity. In practice, some decomposition of the target K_2CrO_4 takes place; however the specific activity of the trivalent ^{51}Cr fraction is much higher than the specific activity achieved directly by the (n, γ) reaction.

In Szilard-Chalmers recoil enrichment processes, the percentage of the total activity of the radionuclide produced which appears in the "recoiled chemical form" is known as the "Szilard-Chalmers recovery" or "recoil enrichment recovery". The percentage of the total activity of the radionuclide which remains in the chemical form of the target is known as the Szilard-Chalmers "retention" (R). The ratio of the specific activities of the recoiled chemical form and that of the target at the end of the irradiation is known as the "enrichment factor" (E). A high percentage recovery and high enrichment factor would mean a successful Szilard-Chalmers process for the production of the radionuclide with high specific activity.

In general, Szilard-Chalmers processes are successful under the following conditions.

1. The target should be stable and should not undergo rapid decomposition under the influence of gamma radiations and fast neutrons.
2. Irradiations should be limited to short periods (a few hours to a few days) and to low and medium neutron fluxes (10^{11}-10^{13}n/cm^2/sec). Irradiations at higher fluxes and for longer durations decompose the target thus reducing the specific activity of the recoiled fraction. The percentage recovery is also reduced under these conditions, due to gamma radiation induced "annealing", i.e., reformation of the original chemical form.
3. The retention 'R' increases on heating the irradiated target. Certain chemicals and sometimes pH and other factors also tend to reduce the recovery and enrichment factor.

Table 17 gives details of some successful Szilard-Chalmers processes employed for preparation of high specific activity radionuclides.

IV. CALCULATION OF YIELDS AND SPECIFIC ACTIVITIES

When targets are irradiated in a nuclear reactor, several processes go on concurrently. Neutrons are captured by nuclei of the target element, giving rise, either directly or through an intermediate beta decay, to nuclei of the desired radionuclide. The nuclei of the product radionuclide may themselves be destroyed by the capture of additional neutrons; they also disappear on account of radioactive decay. The number of target nuclei present is progressively reduced by the primary neutron capture process. The rate of buildup of the required radionuclide is the resultant of all these factors. If no intermediate beta decay is involved or where the half-life of such decay is short, the activity "A" of the product radionuclide is given by the following equation:

$$A = \frac{N_0\, 6_1 \phi\, \lambda_1}{\lambda_1 + 6_2 \phi - 6_1 \phi} \{e^{-6_1 \phi t} - e^{-(\lambda_1 + 6_2 \phi)t}\} \tag{1}$$

Where A is the activity in transformations per second; No is the number of target nuclei originally present; σ_1 is the cross-section of the target nuclei; σ_2 is the cross-section of the product neclei; λ_1 is the decay constant of the product radionuclide; ϕ is the neutron flux; t is the time elapsed since the beginning of the irradiation.

If the period of irradiation is relatively short (as compared to the half-life of the product radionuclide) and the neutron flux is low (less than 10^{12} n/cm^2/sec), the burn-up of the target nuclei can usually be neglected. In most cases, further, the neutron absorption cross-section of the product radionuclide is also small. In such cases, the above equation can be simplified to the well-known form given below.

$$S = \frac{0.6 \times 6 x \phi \times (1 - e^{-0.693t/T})}{3.7 \times 10^{10} \times W} \tag{2}$$

Where S is the specific activity of the product in curies; ϕ is the effective neutron flux in the target (i.e., neutrons/cm^2/sec); 6 is the activation cross-section of the target nuclide in barns; 'W' is the atomic weight of the target element; t is the irradiation time; and T is the half-life of the product radionuclide.

It would be apparent from this equation that the specific activity increases progressively with time of irradiation and reaches a "saturation value" given by

Table 17
SZILARD-CHALMERS PROCESS FOR PREPARATION OF RADIONUCLIDES FOR TRACER APPLICATIONS IN BIOLOGY AND MEDICINE

Radionuclide	Target	Yield	Enrichment	Ref.
^{82}Br	KBrO$_3$	91%	2 × 10^4	Cobble, J. and Boyd, G. E., *J. Am. Chem. Soc.*, 74, 1282, 1952; Boyd, G. E., Cobble, J., and Wexler, S., *J. Am. Chem. Soc.*, 74, 237, 1952
^{82}Br	p-dibromo benzene	6—40%	20—360	Mani, R. S., *Curr. Sci. (India)*, 35, 305, 1966.
^{51}Cr	K$_2$CrO$_4$	5—40%	150—2500	Mani, R. S. and Chowdhary, S. Y., *Curr. Sci. (India)*, 35, 230, 1966
	Acetyl acetonate	76%	260	Toerkoe, J., *Kernenergie*, 7, 753, 1964.
^{64}Cu	Phthalocyanine	40%	1000—3000	Douis, M. and Valade, J., CEA-2072, Report of Commissariat a lenergie atomique, Saclay, France, 1961.
^{64}Cu	Acetylacetonate	40—80%	720	Lin, T. K. and Yeh, S. J., *J. Nucl. Sci. Technol.*, 3, 289, 1966.
^{59}Fe	K$_4$Fe(CN)$_6$ 3H$_2$O	56—67%	2000—9000	Kimura, K., Shibata, N., Yoshihara, K., Tanaka, K., and Nakamura, H., *Proc. 4th Jpn. Conf. Radioisotopes*, Tokyo, 1961, 499.
	Phthalocyanine	44—45%	3	Payne, B. R., Scargill, P., and Cook, G. B., *Radioisotopes in Scientific Research*, Pergamon Press, London, 1957.
^{65}Zn ^{69m}Zn	Phthalocyanine	36—67%	5—120	Payne, B. R., Scargill, P., and Cook, G. B., *Radioisotopes in Scientific Research*, Permanon Press, London, 1957.
^{65}Zn ^{69}Zn	Phthalocyanine		50	Herr, W., *Z. Naturf.*, 76, 201, 1952.

$$S_{sat} = \frac{0.6 \times 6 \times \phi}{3.7 \times 10^{10} \times W} \qquad (2a)$$

(i.e. t > 4T)

for a relatively long irradiation. Figure 1 shows the buildup of specific activity for different periods of irradiation expressed in terms of the half-life of the product radionuclide. It would also be apparent from Equation 2 that the saturation-specific activity is directly proportional to the neutron flux. As explained earlier, however, Equation 2 is applicable only for relatively short periods of irradiation at low fluxes and where the product nuclide does not have a high neutron absorption cross-section. If these conditions are not fulfilled, the more rigorous Equation 1 should be used. In such cases it will be found that a "saturation value" will not be attained, instead the specific activity will rise to a maximum and will decrease thereafter. Further, the maximum-specific activity calculated from Equation 1 will be significantly different from the value obtained on applying Equation 2 to such a case. In applying Equation 1 or 2 or 2a to actual cases, there are other uncertainties also. For example, the effective neutron flux ϕ in the target may be different from the unperturbed thermal neutron flux in the particular irradiation position if the neutron absorption cross-section of the target is high. For most (n, p) and (n, α) reactions, the effective neutron flux in the target will be the flux of "fast neutrons", i.e., neutrons with energies above the "threshold energy" for the particular reaction. There is also some uncertainty about the value of the activation cross-

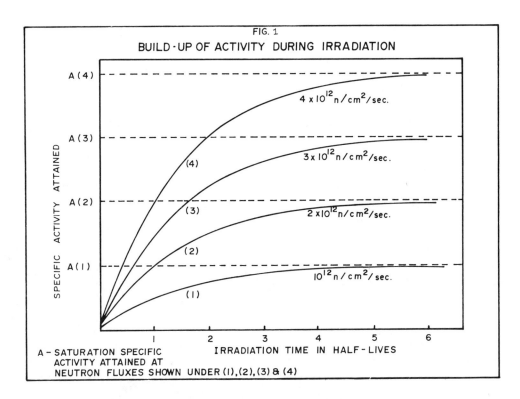

FIGURE 1. Buildup of activity during irradiation.

Table 18
CROSS-SECTIONS OF SOME (n,p) and (n,α) REACTIONS AT DIFFERENT NEUTRON ENERGIES

	Cross-section in millibarns at different neutron energies (MeV)							
	2 MeV	3 MeV	3.5 MeV	4 MeV	5 MeV	6 MeV	7 MeV	14 MeV
^{32}S(n,p)^{32}P	10	140	—	250 (300)	250	320	325	225
^{54}Fe(n,p)^{54}Mn	12	90	—	300	410	490	530	460
^{58}Ni(n,p)^{58}Co	50	140 (200)	—	350	510	600	—	410
^{35}Cl(n,α)^{32}P		12	28	64	—	—	—	—

section σ to be used in the calculation since the cross-sections of target nuclides are dependent on the neutron energy. Table 18 gives the cross-sections of some typical (n, p) and (n, α) reactions at different neutron energies.

In the case of radionuclides which are formed by successive neutron captures, the equations governing the growth of activities are as follows:

$$A_1 = \frac{N_0\, \sigma_1\, \phi\, \lambda_1\, (1 - e^{-\alpha t})}{\alpha} \quad \text{and}$$

$$A_2 = \frac{N_0\, \sigma_1\, \sigma_2\, \phi^2}{\alpha} \left\{ 1 + \frac{\lambda_2}{\alpha - \lambda_2} \cdot e^{-\alpha t} - \frac{\alpha}{\alpha - \lambda_2} \cdot e^{-\lambda_2 t} \right\}$$

where $\alpha = \lambda_1 + 6_2\phi$. λ_2 is the decay constant of the second radionuclide.

Obviously, the production of the second radionuclide will be greater the higher the neutron flux and the longer the duration of irradiation.

V. LARGE-SCALE PRODUCTION CONSIDERATIONS

For undertaking a program for large-scale production and supply of reactor-produced radionuclides, several special facilities, equipment, and materials are required to be procured and installed. The most important among them are (1) reactor irradiation facilities, (2) raw materials including targets for irradiation, special chemicals, and solvents required for processing and purification, (3) handling and processing facilities equipped with gadgets for remote manipulation, (4) facilities for treatment and disposal of radioactive wastes, and (5) set up for sales and marketing of products.

A. Irradiation Facilities

For large-scale production of isotopes, irradiation facilities with neutron fluxes in the range of 10^{13} to 10^{15} n/cm²/sec. are required. Necessary arrangements should be made for rapid target loading and unloading for the production of short-lived radionuclides. Targets are usually enclosed in screw-capped or "cold-welded" pure Al capsules, which are leakproof and can withstand internal pressures of the order of 40 to 50 kg/cm². In many large isotope production reactors, facilities exist for target irradiation inside the reactor core where the highest flux is available, as well as at the periphery where the fluxes are lower. Usually targets for irradiation can be loaded and unloaded without interrupting the reactor operation. Irradiated targets are transported to the processing laboratory in specially designed lead containers; pneumatic transfer facilities are employed for transporting irradiated targets containing very short-lived radionuclides such as ^{18}F.

B. Raw Materials Including Targets

The main raw materials required for production of radionuclides are (1) target chemicals required for reactor irradiation, (2) special chemicals, solvents on ion exchange materials required for chemical processing and purification, and (3) containers and packaging materials.

The target materials employed for reactor irradiation are either pure elements or alloys or stable compounds such as oxides, chlorides, and carbonates. These should be obtained with the required purity and in a form suited for optimum irradiation. For example, Te metal or TeO_2 to be used for irradiation for the production of ^{131}I should have > 99% chemical purity and should contain less than 2 ppm of I, As, Se, Pb, and Sb. Similarly Ni metal to be used for irradiation for production of ^{58}Co should be > 99.5% pure and should contain less than 2 ppm of cobalt impurity. MoO_3 to be used for high flux irradiation for production of ^{99}Mo should be > 99.5% pure and should contain less than 5 ppm of Co, Ag, Sb, Re, Cs, Zr, Cr, and Fe. Sulfur targets required for production of ^{32}P should be > 99.5% pure and should contain less than 2 ppm of C, P, and heavy metals and < 10 ppm of sulfate. Certain targets have to be specially prepared to facilitate postirradiation processing; for example, TeO_2 used for irradiation is required to be purified by sintering at 700°C followed by regrinding. Electromagnetically enriched isotopes are to be employed in some cases to enhance the specific activity of the product radionuclide and to minimize the formation of radionuclidic impurities. For example, $^{58}Fe_2O_3$ (80% enrichment) is used for the production of high specific activity ^{59}Fe containing <1% ^{55}Fe. Enriched ^{235}U (^{235}U > 90%) is employed for the production of ^{99}Mo.

Many special chemicals including high purity acids, solvents, adsorbants for ion exchange and chromatographic separations are required for the processing and purification of reactor-produced radionuclides. These are available with the requisite degree of purity from only a few manufacturers, while in some cases further purification is necessary before these can be used for processing. It is therefore essential to procure them in bulk and carry out further purification and quality control analysis before use. Containers such as glass and polyethylene vials and bottles, shielding containers made of lead or mild steel and packaging materials are required in large quantities for the production program. The primary containers of glass or polyethylene should be carefully selected and cleaned to ensure that they are sturdy and leakproof and do not release impurities.

C. Handling and Processing Facilities

The processing and purification of reactor-produced radionuclides has to be carried out in a Class I radionuclide laboratory. Such a laboratory has to be specially designed with the requisite facilities and services to enable the safe handling of multicurie quantities of radioactive isotopes. The essential requirements for the laboratory are (1) controlled ventilation, (2) air conditioning, (3) shielded and remotely operated "hot cells" and production plants designed and maintained for the production, dispensing, and quality control analysis of radionuclide preparations, (4) facilities for collection, analysis, and storage of radioactive liquid effluents and solid wastes, (5) equipment and services for monitoring, assay, and control of radiation hazards and personnel contamination.

1. Ventilation System

The radionuclide laboratory should be provided with controlled ventilation, ensuring at least 10 changes of air per hour in the high active areas; the entire laboratory should be maintained at a slight negative pressure with respect to the surroundings. The laboratory air should be vented through fume hoods into a tall stack through absolute filters. Arrangement should be made for air monitoring at various locations in the laboratory and its surroundings as well as at the discharge end of the stack, to ensure that there is no release of radioactive material from the laboratory and that the stack discharges are within the permissible levels.

2. Air Conditioning

The radionuclide production laboratory should be air conditioned to ensure reliable functioning of the electronic and other equipment and control instrumentation and for comfort of the working staff.

3. Remote Handling Facilities

The main remote handling facilities required for large-scale production of radionuclides are "master-slave manipulator" and "mini manipulator" cells, lead shielded production, and dispensing plants fitted with tongs, can-cutters, remote pipetting units, and control instrumentation including beta-gamma ionization chambers, pH meters, and vial sealing machines. The master slave and mini-manipulator cells are generally constructed of concrete, with stainless steel lining and are equipped with master-slave or mini manipulators having electromechanical controls. The lead shielded production plants generally consist of sealed aluminum or mild-steel or plastic boxes with provision of gaiters for tongs, can cutting machines, pipetting units, etc. For these boxes, shielding with lead "bricks" is provided on all sides as well as on top and bottom. The manipulator cells as well as production plants are maintained at a slight negative pressure with respect to the laboratory and have the requisite number of air changes per hour; they are also equipped with radiation monitoring and survey instrumentation coupled to automatic alarm systems. Transfer ports are provided for these cells and plants for inflow and outflow of materials and for facilitating cleaning and maintenance operations; such ports are generally provided with safety interlocks to ensure that they are not fully opened when the radiation level inside is higher than permissible. In addition, the laboratory should be equipped with the necessary number of radiochemical fume hoods, having the requisite air flow.

4. Effluent, Waste Collection, and Storage

The radionuclide laboratory should have active sinks and underground high and low active drains leading to "sump" and "delay" tanks for collection, hold up, and storage of radioactive effluents. Provision should be made in these tanks for sampling of the effluents for analysis, for addition of chemicals for pH control, precipitation, etc., and for pumping out the liquid. The laboratory should also be equipped with receptacles and shielded "vaults" for storage of active solid wastes. In the collection and storage of wastes, long-lived and hazardous wastes containing high levels of activity should be clearly segregated from short-lived isotopes and low level wastes.

5. Control of Radiation Hazards

A vigilant health physics group should be organized to look after radiological health protection of the working personnel. This group should be responsible for environmental contamination control in the production laboratories, for material and personnel decontamination, for maintenance of protective equipment, measurement, and recording personnel exposures, and for investigation and prevention of accidents.

D. Treatment and Disposal of Wastes

In undertaking a program for large-scale production of reactor-produced radionuclides, it will be necessary to set up an ancillary facility which will carry out the following functions: (1) treatment and disposal of low and medium level liquid effluents: (ion exchange and fixation on natural siliceous materials may be employed for this purpose), (2) solidification and immobilization of long-lived wastes, (3) incineration of low level flammable wastes, solid and liquid, and (4) storage and management of high level wastes, both liquid and solid.

E. Sales and Marketing

A large manufacturer of reactor-produced radionuclides should have an efficient sales and marketing set up for prompt customer service and for efficient execution of sales and supplies.

F. Other Requirements for Large-Scale Production

A large-scale manufacturer and supplier of reactor-produced radionuclides should have facilities for efficient maintenance and servicing of the remote processing equipment, production plants, electronic counting and standardization equipment and of laboratory instruments. An efficient operations and maintenance (0 & m) cell should be constituted for this purpose and for looking after the laboratory ventilation and air conditioning. Special attention must be paid to prevent corrosion of the equipment used inside the production plants. The performance of automatic in-process control equipment such as pH meters, ionization chambers, remote dispensing, weighing, measuring, and vial sealing machines should be carefully checked from time to time.

Proper records must be maintained for all production batches. Control samples should be maintained for all batches, so that these are available for examination in case of complaints. Adequate provision should be made for the proper storage of all stocks and control samples.

The personnel employed for the production, dispensing, quality control, and other operations should be suitably trained and organized for maximum internal coordination, efficiency, and objectivity in quality control.

VI. TYPICAL PRODUCTION PROCESSES

A few typical production processes employed for large-scale production of some important reactor-produced radionuclides are briefly described below.

A. Tritium

Tritium is produced on a large-scale by the irradiation of natural *lithium* or enriched *lithium-6* usually in the form of an alloy with magnesium or aluminum.[21-23]

After irradiation, the alloy is heated to near its melting point in vacuum when tritium and other gases are released. The hydrogen isotopes are collected by adsorption on pyrophoric U. Further purification of H_2-T_2 mixture is carried out by gas chromatography using palladium membrane. For obtaining more than 99% isotopic purity the thermal diffusion process is employed.

B. Carbon-14

Beryllium nitride and aluminum nitride are the targets of choice. Wet and dry distillation methods have been adopted for the separation of ^{14}C from the irradiated targets [24,25,26] In the dry method[25] irradiated Al N is mixed with decarbonized CuO and is heated to 1000°C in a current of decarbonized O_2. The $^{14}CO_2$ produced is trapped in saturated $Ba(OH)_2$ solution. The precipitated $Ba^{14}CO_3$ is filtered, washed, dried and assayed. In the wet distillation process[24, 26] Al N or Be_3N_2 is dissolved in H_2SO_4 or H_3PO_4 and the $^{14}CO_2$ produced is flushed out and trapped in saturated $Ba(OH)_2$ solution. The untrapped $^{14}CH_4$ originating from the hydrolysis of carbide is oxidized by passing through decarbonized CuO held at a temperature exceeding 700°C. The resulting $^{14}CO_2$ is again trapped in $Ba(OH)_2$. The residual ^{14}C remaining in the dissolved target is oxidized with $K_2Cr_2O_7$ and H_2SO_4 and the $^{14}CO_2$ is trapped again in $Ba(OH)_2$. Decarbonized O_2 gas is used to flush out the $^{14}CO_2$. The precipitated $Ba^{14}CO_3$ is washed, dried and assayed. To obtain the highest specific activity for ^{14}C, the target should contain <30 ppm of carbon impurity. The target is usually irradiated at a neutron flux exceeding 10^{13}n/cm^2/sec for 2 years.

C. Sodium-24

Sodium-24 is prepared by the nuclear reaction ^{23}Na (n, γ) ^{24}Na. Sodium carbonate or bicarbonate is the target employed for production. Irradiation is carried out at neutron fluxes of the order of 10^{13} n/cm^2/sec for several hours. The irradiated carbonate or bicarbonate is dissolved in dilute HCl and the solution is boiled to remove CO_2. After cooling and pH adjustment to 7-8 (by adding dilute NaOH) the solution is diluted to obtain 0.9% NaCl solution. The solution is dispensed into vials and is sterilized. The above operations are carried out by remote control using simple glass ware.

The product sodium chloride-^{24}Na has the following specifications:

Chemical form	Colorless aqueous solution of sodium chloride, isotonic
Radioactive concentration	1 — 3 m Ci/mℓ
Specific activity	200 — 1000 m Ci/g Na

D. Potassium-42 and Rubidium-86

Potassium-42 and Rubidium-86 are similarly processed from irradiated K_2CO_3 and Rb_2CO_3, respectively. *Calcium-45* is prepared from irradiated $CaCO_3$; however, electromagnetically enriched $^{44}Ca\ CO_3$ (natural abundance 2.09%) may be employed to obtain ^{45}Ca of higher specific activity.

Specifications of potassium chloride-^{42}K:

Chemical form	Colorless aqueous solution of KCl, isotonic
Radioactive concentration	0.1—1 m Ci/mℓ
Specific activity	20—500 m Ci/g K

Specifications of rubidium chloride-^{86}Rb:

Chemical form	RbCl in aqueous solution, isotonic
Radioactive concentration	0.5—5 m Ci/mℓ
Specific activity	50—300 m Ci/g Rb

Specifications of calcium chloride-^{45}Ca:
  Chemical form   CaCl$_2$ in dilute HCl solution
  Radioactive concentration 1—5 m Ci/mℓ
  Specific activity   10—500 m Ci/g Ca

E. Phosphorus-32

Pure elemental S is the target employed. Irradiation is carried out for 4 to 6 weeks at the highest fast neutron flux available in the core of the reactor. ^{32}P is recovered from the irradiated S by distilling off the S either at reduced pressure or in a stream of N$_2$. Distillation at reduced pressure ($\sim$1 mmHg) can be carried out at a temperature of 120 to 150°C, while distillation with N$_2$ as carrier gas is carried out at about 500°C. After complete distillation of S, the ^{32}P which is left behind in the distillation container is leached out with dilute HCl and is purified by passage through a cation exchange column. The distillation of S is carried out in a silica still; however, reduced pressure distillation may be carried out in an all-glass container.[27-30]

Specifications of ortho phosphoric acid-^{32}P
  Chemical form   Colorless dilute HCl solution of ortho phosphoric acid
  Radioactive concentration 5—50 m Ci/mℓ
  Radiochemical purity   >99% as O-phosphoric acid
  Specific activity   >25 Ci/mg P

F. Sulfur-35

Pure KCl is the target used, and irradiation is carried out for 3 to 6 months at a high fast neutron flux. After a postirradiation cooling of 1 week for ^{42}K and ^{38}Cl to decay out, the target is processed to recover the ^{35}S.

For large-scale production of ^{35}S, 2 main methods have been employed. The first method is based on cation exchange removal of K$^+$ from irradiated KCl solution, followed by distillation of the chloride as HCl.[31] The second method is based on selective adsorption of HSO$_4^-$ on alumina[32] from a dilute HCl solution of the irradiated KCl. The HSO$_4^-$ retained on the alumina is eluted with dilute ammonia and any traces of Al are removed either by cation exchange separation[32] or by precipitation as hydroxide.[33]

Specifications of sulfuric acid-^{35}S:
  Chemical form   H$_2$SO$_4$ in dilute HCl solution
  Radioactive concentration >10 mCi/mℓ
  Specific activity   >10 Ci/mg S

G. Chromium-51

Two methods are employed for the production of ^{51}Cr of high specific activity, the first based on direct (n, γ) activation of electromagnetically enriched ^{50}Cr (natural abundance 4·35%) and the second is based on Szilard-Chalmers enrichment using K$_2$CrO$_4$ as target.[17,18]

1. (n, γ) Activation of Enriched ^{50}Cr

Electromagnetically enriched ^{50}Cr (80 to 90%) in the form of metal or oxide is irradiated at a high neutron flux (10^{13} to 10^{14} n/cm^2/sec) for about 8 weeks. The irradiated target is dissolved in HCl and the chromic chloride is oxidized to chromate.

2. Szilard-Chalmers Enrichment

K$_2$CrO$_4$ is the most commonly used target, and is irradiated at neutron fluxes of 3×10^{12} to 10^{13}n/cm^2/ sec for 3 to 5 days. The enriched ^{51}Cr^{III} may be separated from the target K$_2$CrO$_4$ by direct precipitation as hydroxide or by carrier precipitation using Fe, or Al or La carrier. The separated ^{51}Cr^{III} hydroxide is oxidized to chromate and the carrier (if any) is removed by precipitation or solvent extraction.

The product sodium chromate-^{51}Cr has the following specifications:

Chemical form	Sodium chromate in isotonic solution
Specific activity	50—1500 mCi/mg Cr
Radioactive concentration	1—5 mCi/mℓ
Radiochemical purity	Greater than 98% of the ^{51}Cr is present as chromate

The Szilard-Chalmers enrichment method can be adopted for production of high specific activity ^{51}Cr using low to medium flux irradiation facilities.

H. Cobalt-58

Very pure nickel metal containing less than 2 ppm of cobalt impurity is the target. Irradiation is carried out at a high fast neutron flux for 3 to 6 months. The irradiated Ni is dissolved in HNO_3 and HCl and is converted to the chloride in 9 M HCl solution. This solution is passed down an anion exchanger which retains the cobalt in the form of a chloro complex. The nickel is removed completely by washing wtih 9 M HCl and the cobalt is eluted with 4-5 M HCl.

Specifications of the cobalt chloride-^{58}Co:

Chemical form	Co Cl_2 in dilute HCl solution
Specific activity	> 3000 Ci/g Co
Radioactive concentration	10—20 mCi/mℓ > 99.5%
Radionuclidic purity	^{60}Co content < 0.1%

I. Iron-59

Electromagnetically enriched ^{58}Fe as oxide, containing not less than 80% ^{58}Fe, is irradiated for 3 to 6 months at the highest neutron flux available. The irradiated target is dissolved in HCl, evaporated to dryness and then taken with dilute HCl.

Specifications of ferric chloride-^{59}Fe:

Chemical form	$FeCl_3$ in dilute HCl solution
Specific activity	10—50 m Ci/mg Fe
Radioactive concentration	5—20 m Ci/mℓ >99.9%, excluding ^{55}Fe
Radionuclidic purity	^{55}Fe < 0.5%
	^{60}Co < 0.001%

J. Molybdenum-99

Two methods are employed for the production of ^{99}Mo of high specific activity, which is mainly used for the preparation of ^{99m}Tc generators, namely, (1) direct irradiation of molybdenum oxide, and (2) fission of uranium (natural or enriched).

1. Direct Irradiation of MoO_3

Pure MoO_3 is irradiated at neutron fluxes exceeding 2×10^{14} n/cm^2/sec for 5 to 7 days, and is then dissolved in NH_4OH. To enhance the specific activity of ^{99}Mo, enriched ^{98}Mo O_3 is used in some cases.

The ^{99}Mo solution obtained has the following specifications

Chemical form	Ammonium molybdate solution
Specific activity	Greater than 5 Ci/g
Radioactive concentration	25—100 m Ci/mℓ
Radionuclidic purity (excluding ^{99m}Tc	> 99.5% (see Table 5)

2. Fission of Uranium

Natural or enriched ^{235}U in the form of metal or U-Al alloy or uranium oxide is the target usually employed. In a typical production, an alloy of U and Al, ^{235}U enriched to 90% is irradiated at a neutron flux of 2×10^{13} n/cm^2/sec for 3 to 7 days. The yield of ^{99}Mo is approximately 15 Ci/g ^{235}U. The irradiated target is dissolved in HNO$_3$ with Hg(NO$_3$)$_2$ as a catalyst. A small amount (about 0.5 mg) of Te carrier is added (in HNO$_3$ solution) and after appropriate dilution, the solution is passed down an Al$_2$O$_3$ column. The ^{99}Mo is adsorbed on the column which is then washed with 1 M HNO$_3$ to remove U and fission product contaminants. The column is then washed with water and gradient elution with dilute NH$_4$OH gives ^{99}Mo free from all radionuclidic contaminants. The ^{99}Mo thus obtained has the following specifications

Chemical form	Ammonium molybdate solution
Specific activity	Greater than 50 Ci/mg
Radioactive concentration	100—300 m Ci/mℓ
Radionuclidic purity (excluding ^{99m}Tc)	> 99.9%
	^{131}I < 0.025 µCi/mCi
	^{103}Ru < 0.025 µCi/m Ci
	^{89}Sr < 0.0003 µ Ci/m Ci
	^{90}Sr < 0.00003 µ Ci/m Ci
All other beta and gamma emitting impurities	< 0.05 µCi/m Ci
Gross alpha activity	< 0.0005 nCi/m Ci

K. Iodine-125

Natural xenon containing 0.096% ^{124}Xe is the target. In a typical production, the Xe gas is filled under pressure in a Zircaloy-2 capsule at 140 kg/cm^2 pressure. Irradiation is carried out at a neutron flux of 2×10^{14} n/cm^2/sec for 30 to 60 days. The yield of ^{125}I is approximately 0.2 Ci/g Xe. The irradiated Xe is allowed to decay for about 60 days to reduce the ^{126}I level to less than 2%. The irradiated capsule is cooled to liquid nitrogen temperature and then opened and the temperature is raised to sublime off the Xe gas. After complete removal of Xe gas, the capsule is again cooled and the ^{125}I is leached out with dilute NaOH solution. The solution, after pH adjustment and filtration is passed through a cation exchange column to remove cationic impurities and the traces of ^{137}Cs produced by the ^{136}Xe (n, γ) $^{137}_{54}$Xe $\xrightarrow{\beta}$ $^{137}_{55}$Cs reaction.[34]

The ^{125}I solution has the following specifications:

Chemical form	Na I in dilute NaOH solution
Specific activity	Greater than 15 Ci/mg
Radioactive concentration	100—300 m Ci/mℓ
Radionuclidic purity	Greater than 99.9% exclusive of ^{126}I, ^{126}I < 1%
Radiochemical purity	Greater than 98% as I$^-$

L. Iodine-131

The method of production employed in most production centers now is by irradiation of Te metal or TeO$_2$. Production from irradiated U which was employed earlier at some centers has been abandoned on account of the need to have elaborate equipment and waste handling facilities.

Both wet and dry distillation methods are employed to recover ^{131}I from irradiated Te targets.

1. Wet Distillation Methods

Irradiated Te metal is dissolved in chromic acid — H$_2$SO$_4$ mixture. After complete dissolution, the iodate formed is reduced to elemental iodine with oxalic acid, and is then distilled off. The distillate is trapped in alkaline sulfite solution. This is then subjected to a second oxidation-reduction cycle using KMnO$_4$ and oxalic acid and iodine is again distilled off and trapped in dilute alkali.[35]

In another wet distillation method, irradiated TeO_2 is dissolved in NaOH and the sodium tellurite is oxidized to tellurate with H_2O_2 in the presence of a catalyst sodium molybdate. The mixture is acidified with H_2SO_4 to 6 to 8 N and the iodine is distilled off and trapped in ice-cold water.[36]

2. Dry Distillation Method

In a typical dry distillation method, purified TeO_2 is irradiated. The irradiated TeO_2 is then heated to 700°C in a stream of air when the iodine distills off and is trapped in dilute NaOH.[37]

3. Relative Advantages and Disadvantages

Many variations of dry and wet distillation methods have been in use for the production of ^{131}I.[35,36] Dry distillation methods employ TeO_2 or Te and involve heating at temperatures in the range of 600 to 800°C. The equipment has to be of quartz. The advantages of dry distillation methods are (1) very little of radioactive waste is generated and (2) since no extraneous reagents are employed, ^{131}I of higher purity and specific activity than is obtainable by wet methods can be produced. The need to work at high temperatures and the high cost of quartz apparatus are the two disadvantages. The advantages of the wet methods are relatively inexpensive equipment and working at lower temperatures. The disadvantages of wet methods are (1) large volumes of liquid wastes are produced, and (2) substantial quantities of extraneous chemicals and reagents are to be employed; this may result in some dilution of the specific activity of the ^{131}I product on account of carry-over of traces of iodine impurity from these chemicals.

Specifications of ^{131}I

Chemical form	NaI in dilute NaOH solution
Specific activity	~ 20 Ci/mg I
Radioactive concentration	> 50 mci/mℓ
Radionuclidic purity	> 99.9%. No impurity is detected by gamma spectrometry
Radiochemical purity	> 98% of ^{131}I is in the form of iodide
Chemical purity	As, Pb, Se <5 μg/mℓ
	Te <2 μg/mℓ
	Other metals <1 μg/mℓ
	Reducing agents <5 × 10^{-4}N agents as sulfite

Sodium iodide ^{131}I of high specific activity in aqueous or dilute NaOH solution undergoes self-radiolysis and oxidation to iodate and other chemical forms. Addition of $Na_2S_2O_3$ prevents this and preserves the ^{131}I in the form of I^-. The recommended concentration of $Na_2S_2O_3$ is 0.05 to 0.1 mg/m Ci ^{131}I.[16] Thiosulfate, however, interferes in the use of ^{131}I for labeling organic compounds, since the labeling process necessarily involves the conversion of I^- to elemental iodine or I Cl which is inhibited by sulfite, thiosulfate, oxalate, and other reducing agents.[14]

VII. QUALITY CONTROLS

Quality controls of reactor produced radionuclide preparations consist of the following:

A. Evaluation of Radionuclidic Purity

This is generally carried out by gamma spectrometry using either NaI(Tl) or Ge(Li) crystal detectors or by beta ray absorption measurements and beta ray scintillation spectrometry. In a few cases, specific chemical separation of the radionuclide, followed by assay is employed. For example, traces of ^{24}Na impurity present in ^{42}K can be evaluated by NaI(Tl) gamma spectrometry; the 2.754 MeV gamma photopeak of ^{24}Na can be measured to estimate the percentage of the impurity. Traces of ^{60}Co impurity in ^{58}Co can be evaluated by gamma spectrometry using a Ge(Li) detector; the 1.17 MeV gamma photopeak of ^{60}Co can be measured to estimate the percentage of the impurity. Beta ray absorption measurements can be used to determine ^{32}P impurity in ^{35}S. Traces of fission product impurities such as ^{131}I, ^{103}Ru, and ^{90}Sr as well as alpha emitting impurities are estimated in fission produced ^{99}Mo by selective extraction of the bulk of ^{99}Mo after adding molybdate carrier and hold-back carriers of sodium iodide and strontium nitrate. For extraction, molybdenum can be converted to silico molybdate heteropolyacid from acid nitrate solutions and then extracted with butanol.[38]

B. Evaluation of Radiochemical Purity

Radiochemical purity is evaluated by paper, column, or thin-layer chromatography or paper electrophoresis, followed by counting. For example, the radiochemical purity of ^{32}P in the form of ortho phosphoric acid is determined by paper chromatography using a solvent mixture consisting of isopropanol, ammonia, and trichloro acetic acid.[9,10] In this system, the Rf of o-phosphate is approximately 0.76, that of meta phosphate approximately 0.00 and of pyrophosphate approximately 0.40.

Table 19 gives brief details of the methods employed for evaluation of the radiochemical purity of some important reactor-produced radionuclides.

C. Evaluation of Chemical Purity

Undesirable metallic impurities which might be present in preparations of reactor-produced radionuclides are estimated semiquantitatively by spot tests or more quantitatively by emission spectrography. Hydroxide and sulfide precipitation tests are employed to estimate heavy metal impurities. Reducing agent impurities in solutions of iodine-131 are estimated by microtitration using standard potassium permanganate and oxalic acid solutions.

D. Other Aspects

Other aspects of quality control of reactor-produced radionuclide preparations include estimation of specific activity and the concentration of vehicles and any additives. For estimation of the specific activity, colorimetric methods are generally employed. For example, the absolute concentration or chromate in solutions of sodium chromate-^{51}Cr is determined spectrophotometrically,[8,10] while the concentration of iodide in sodium iodide-^{131}I is determined spectrophotometrically[39] or by using a specific iodide electrode.[40] Simple volumetric or gravimetric methods are employed to measure the concentrations of additives. For example, the concentration of sodium chloride in sodium chromate-^{51}Cr solution can be estimated by titration with standard silver nitrate solution.

VIII. PACKAGING AND TRANSPORT

Radionuclides are packed for transport in conformity with regulations drawn up by the IAEA. The basic requirements for the shipping containers are (1) safe containment of the contents, (2) protection of personnel and certain objects such as unexposed photographic films against radiation, and (3) economy in transport.

The radionuclide preparation is generally enclosed in a primary container that is leakproof and securely sealed and is then enclosed in a shielding container adequate to prevent an external radiation dose rate in excess of certain limit specified below. For radioactive preparations in liquid form, there is the additional requirement that the vessel should have free space to prevent the buildup of gas pressure due, for example, to radiolysis of the solvent. Further, the primary containers in these cases should be surrounded by enough absorbent material to absorb the entire contents in case of an accidental breakage.

Currently, two sets of regulations for the transport of radioactive materials, one based on the 1967 edition of the IAEA regulations for the safe transport of radioactive materials and

Table 19
EVALUATION OF RADIOCHEMICAL PURITY OF IMPORTANT REACTOR-PRODUCED RADIONUCLIDES

Radionuclide and chemical form	Radiochemical impurity	Method of analysis and reference
^{18}F NaF in dilute NaOH	AlF_6^{3-}	Paper chromatography using the solvent mixture butanol-methanol-water 1:3:1 Rf F$^-$ 0.53^{50}
^{32}P H_3PO_4 in dilute HCl	meta, pyro and poly phosphoric acids	Paper chromatography using the solvent isopropanol 75 mℓ, water 25 mℓ, trichloroacetic acid 5 g, ammonia 0.3 mℓ or n-propanol 30 mℓ, ethanol 30 mℓ, water 30 mℓ, ammonia 1 mℓ, Rf o-phosphate 0.76, meta 0.00, pyro 0.40.^{9,10}
^{51}Cr Na_2CrO_4 in NaCl solution	$^{3+}$Cr Colloidal $Cr(OH)_3$	(i) $^{3+}$Cr is separated from chromate by anion exchange, after adding carrier chromate and chromic nitrate or chloride.^{9} (ii) Paper chromatography using the solvent ethanol, ammonia, water 50:25:125, Rf of $CrO_4^=$ 0.9, Rf of Cr^{3+} 0.00.^{10}
^{51}Cr $CrCl_3$ in dilute HCl	CrO_4^-	Paper chromatography using the solvent ethanol, ammonia, water (50:25:125) Rf of $^{3+}$Cr = 0.00, Rf of $CrO_4^=$ = 0.90.^{10}
^{125}I ^{131}I NaI in dilute NaOH or NaI in dilute $Na_2S_2O_3$ or sulphite solution	IO_3^-, IO_4^-	(i) Paper chromatography using the solvent 75% aqueous methyl alcohol. Rf I$^-$ ~ 1.0, Rf IO_3^- ~ 0.46, Rf IO_4^- ~ 0.00.^{10} (ii) Paper electrophoresis in 0.05 M Na_2HPO_4 solution whatman No. 540 paper at 8V/cm for 21/2 hours.^{10}
^{99m}Tc sodium pertechnetate in dilute NaCl solution	Reduced ^{99m}Tc species	Paper chromatography using the solvent mixture 85 parts methanol and 15 parts water. Rf of TcO_4^- 0.5—0.6, Rf of reduced Tc species 0.00.^{51}

the other based on the 1973 revised edition of these regulations are in vogue. Either set of regulations may be adopted, provided the same is acceptable to the governments of the countries of origin, transit and destination of the radioactive consignment. Both sets of regulations are complete in themselves in so far as radioactive materials are concerned, and salient features of these regulations are briefly summarized below.[41]

A. IAEA Regulations of 1967

As per these regulations, the different radionuclides are divided into seven groups on the basis of their radiotoxicity:

- Group I—Highly toxic consisting of mainly long-lived alpha emitters such as ^{239}Pu, ^{226}Ra, and ^{210}Po.
- Group II—Short-lived α and long-lived β$^-$ emitters such as ^{233}Pa, ^{223}Ra, ^{210}Pb, and ^{90}Sr.

- Groups III and IV—Short to medium half-life radionuclides of low toxicity.
- Group V—Uncompressed ^{85}Kr, ^{41}Ar, etc.
- Group VI—^{133}Xe (uncompressed).
- Group VII—Tritium (compressed or uncompressed gas) and its sources.

For routine shipment two packaging types are used, namely, Type A, which is intended to withstand ordinary conditions of carriage and handling, including minor accidents, and Type B, which is intended to withstand ordinary conditions of carriage and severe accidents. The maximum amounts of radionuclides of Groups III and IV which can be transported in Type A containers range from 3 Ci for ^{125}I, ^{131}I, and ^{133}Xe, to 20 Ci for ^{14}C, ^{99}Mo, ^{57}Co, ^{51}Cr, ^{32}P, etc.; for ^{85}Kr and ^{3}H the maximum quantities allowed are 1000 Ci.

It is apparent that the large majority of reactor-produced radionuclides which are of interest as tracers in the biomedical sciences can be shipped in Type A containers.

Type A containers meant for solid or liquid preparations must withstand certain tests meant to confirm that these packages are not seriously damaged by rain, by dropping them from a height, or by compression. Depending on the external radiation dose, the packages again are categorized into "white" and "yellow" as detailed below:

1. Category I: White signifies that the dose rate at the surface of the package does not exceed 0.5 mR/hr or equivalent. The most sensitive photographic film can stand 20 hr of intimate contact with such a package without fogging.
2. Category II: Yellow signifies that the dose rate of radiation originating from the package shall not exceed, at any time during transportation (a) 50 mR/hr or equivalent at any point on the external surface of the package; and (b) 1 mR/hr or equivalent at a distance of 1 meter (40 in) from the external surface of the package. Further the 'Transport Index' (defined as the number expressing the maximum radiation level in millirem per hour at 1 m from the external surface of the package) shall not exceed 1 at any time during transportation.
3. Category III: Yellow signifies that the dose rate of radiation originating from the package shall not exceed, at any time during transportation: (a) 200 mR/hr or equivalent at any point on the external surface of the package and (b) 10 mR/hr or equivalent at 1 m from the external surface of the package.

Further, the Transport-Index shall not exceed 10 at any time during transportation. The levels of radiation emitted from the package are thus indicated by the labels that are attached.

The labels should also indicate the principal radioactive content and the total activity in curies or millicuries. The sum of the transport indices of the packages in one transport vehicle should not exceed 50.

Small quantities of radioactive materials are classed as "exempt"; in these cases the containers need not meet the requirements of Type A packages, though these must be leakproof and satisfactory to contain the material safely. The external radiation at the surface of such packages should not exceed 0.5 mR/hr. The maximum "exempt" quantities of radionuclides are 1 mCi for preparations of isotopes of Groups III to VI and 25 Ci for isotopes of Group VII. The inner container must, however, be clearly marked "Radioactive" in the packages.

B. 1973 IAEA Regulations

Two types of packages, Type A and Type B, are specified. Type A packaging shall mean a packaging that is designed to withstand the normal conditions of transport as demonstrated by the retention of the integrity of containment and shielding after certain specified tests. Type B packaging shall mean a packaging that is designed to withstand the damaging effects of transport accident as demonstrated by the retention of the integrity of containment and shielding after certain specified tests. Type A is again classified into special form radioactive material — A1 and others and A2. Special form radioactive material shall mean either an indispersible solid radioactive material or a sealed capsule containing the radioactive material. The sealed capsule shall be so constructed that it can be opened only by destroying the capsule. A Type A package, since its contents are limited to A1 or A2 does not require competent authority approval. The A1 and A2 values for some of the important radionuclides of interest as tracers in biology and medicine are given in Table 20.

Type B packages are again classified as Type B (M) and Type B (U). Type B (U) package requires unilateral approval only (approval of the competent authority of the country of origin) while Type B (M) package requires multilateral approval (approval of the competent authority of the country of origin and of each country through or into which the consignment is to be transported).

The packing, labeling, and marking specifications are generally in accordance with the specifications detailed under the 1967 IAEA regulations. Both sets of regulations have stipulated the requirements of shipping documents, shippers certification, and handling and loading.

IX. CURRENT AVAILABILITY OF REACTOR-PRODUCED RADIONUCLIDES

Commercial producers in many advanced countries offer a wide range of reactor-produced radionuclides with specifications and characteristics suited for tracer applications. In addition to local supplies, many of them undertake overseas shipments. The national atomic energy organizations in many countries have undertaken production and supply of radionuclides as part of their program of development and utilization of atomic energy. In many instances, such production is mainly confined to radionuclides of relatively short half-life ranging from a few hours to a few days, and is supplemented by imports, particularly of long-lived products.

A large-scale producer having high flux irradiation facilities and a well-organized processing and marketing set up has several advantages as against a small-scale producer. (1) Irradiations at high fluxes give radionuclides of high specific activities which are often required for many tracer applications. Such high specific activities are often difficult to attain in low flux irradiations. (2) It is possible to effect the maximum economy in the use of precious targets, consisting of enriched isotopes, in high flux irradiations. (3) A large-scale producer has the possibility of holding stocks, even of relatively short-lived radionuclides. (4) A well-organized quality assurance and quality control program can be instituted more economically in large-scale production than in small-scale preparation. However, the difficulties in regular procurement of very short-lived radionuclides from overseas and national policies in the field of atomic energy development have provided the justification for many countries to take up local production of radionuclides.

Table 21 gives a listing of important large-scale producers and national atomic energy organizations the world over currently processing and supplying reactor-produced radionuclides.

X. FUTURE PERSPECTIVES

During the last three and a half decades following the announcement of the availability of reactor-produced radionuclides in 1946, the potentialities of the nuclear reactor for radionuclide production have been well explored. In the first two decades of this period, the reactor almost completely overshadowed the cyclotron that practically all the developments in the field of tracer applications were confined to reactor-produced radionuclides. However in the early 1960s came the development of many useful cyclotron-produced radionuclides

Table 20
A_1 AND A_2 VALUES FOR REACTOR-PRODUCED RADIONUCLIDES FOR TRACER APPLICATIONS

Symbol of radionuclide	A_1 (Ci) special form	A_2 (Ci) other forms	Symbol of radionuclide	A_1 (Ci) special form	A_2 (Ci) other forms
^{111}Ag	100	100	^{87}Kr (compressed)	0.6	0.6
^{76}As	10	10			
^{77}As	300	300	^{140}La	30	30
^{198}Au	40	40	^{177}Lu	300	300
^{199}Au	200	200	^{28}Mg	6	6
^{131}Ba	40	40	^{56}Mn	5	5
^{133}Ba	40	10	^{99}Mo	100	100
^{140}Ba	20	20	^{24}Na	5	5
^{82}Br	6	6	^{95}Nb	20	20
^{14}C	1000	100	^{32}P	30	30
^{45}Ca	1000	40	^{193m}Pt	200	200
^{47}Ca	20	20	^{197}Pt	300	300
^{109}Cd	1000	70	^{86}Rb	30	30
^{115m}Cd	30	30	^{188}Re	10	10
^{115}Cd	80	80	^{35}S	1000	300
^{36}Cl	300	30	^{122}Sb	30	30
^{38}Cl	10	10	^{124}Sb	5	5
^{58m}Co	1000	1000	^{47}Sc	200	200
^{58}Co	20	20	^{113}Sn	60	60
^{60}Co	7	7	^{85m}Sr	80	80
^{51}Cr	600	600	^{85}Sr	30	30
^{131}Cs	1000	1000	T (uncompressed)	1000	1000
^{137}Cs	30	20			
^{64}Cu	80	80	T (compressed)	1000	1000
^{67}Cu	200	200			
^{18}F	20	20	T (tritiated water)	1000	1000
^{55}Fe	1000	1000			
^{59}Fe	10		T (other forms)	20	20
^{72}Ga	7	7			
^{197m}Hg	200	200	^{96m}Tc	1000	1000
^{197}Hg	200	200	^{97m}Tc	1000	200
^{203}Hg	80	80	^{99m}Tc	100	100
^{125}I	1000	70	^{132}Te	7	7
^{129}I	1000	2	^{131m}Xe (uncompressed)	100	100
^{131}I	40	10			
^{132}I	7	7	^{131m}Xe (compressed)	10	10
^{113m}In	60	60			
^{114m}In	30	20	^{133}Xe (uncompressed)	1000	1000
^{115m}In	100	100			
^{42}K	10	10	^{133}Xe (compressed)	5	5
^{85m}Kr (uncompressed)	100	100			
			^{90}Y	10	10
^{85m}Kr (compressed)	3	3	^{91m}Y	30	30
			^{169}Yb	50	50
^{85}Kr (uncompressed)	1000	1000	^{65}Zn	30	30
			^{69m}Zn	40	40
^{85}Kr (compressed)	5	5	^{69}Zn	300	300
			^{95}Zr	20	20
^{87}Kr (uncompressed)	20	20	^{91}Y	30	30

Table 21
IMPORTANT LARGE-SCALE MANUFACTURERS OF REACTOR-PRODUCED RADIONUCLIDES AND THEIR PREPARATIONS

United States of America

1. General Electric, Pleasanton, California
2. Union Carbide International Co., 270 Park Avenue, New York
3. New England Nuclear 575, Albany Street, Boston, Massachusetts[a]
4. Mallinckrodt Nuclear, Mallinckrodt Chemical Works, St. Louis, Missouri[a]
5. E. R. Squibb and Sons, Inc., New Brunswick, New Jersey[a]
6. Abbot Laboratories, North Chicago, Illinois[a]
7. Becton Dickinson Immunodiagnostics, Orangeburg, New York[b]
8. Ames Company, Division of Miles Laboratories, Inc., Elkhart, Indiana[b]
9. Clinical Assays, 620 Memorial Drive, Cambridge, Massachusetts[b]
10. Cambridge Nuclear 575, Middlesex Turnpike, Billerica, Massachusetts[b]
11. Radioassay Systems Laboratories, 1511, East Del Amo Blvd., Carson, California[b]
12. Calbiochem, 10933 N. Torrey Pines Road, La Jolla, California[b]
13. Diagnostic Products Corporation, 12306, Exposition Boulevard, Los Angeles, California[b]
14. Nuclear Medical Systems, 1531, Monrovia Avenue, Newport Beach, California[b]

Canada

1. Charles E. Frost and Co., P. O. Box 24, Montreal 3, Quebec[a]
2. Diagnostic Biochem Canada, Inc., 249, Wortley Road, London, Ontario, Canada[b]
3. Bio-RIA, 10900, Hamon, Montreal, Canada H3M 3A2[b]

Europe

1. The Radiochemical Centre, Amersham Buckinghamshire, England
2. Commissariat A Lenergie Atomique Saclay, France
3. Sorin Biomedica, Saluggia, Italy
4. Institut National des Radioelements, 6220, Fleurus, Belgium
5. Isotope Service, Aktiebolaget, Atomenergi, Studsinkfack, Sweden
6. Farbwerke, Hoechst AG, Venkauf, Arzneimittel, 6230, Frankfurt (M), 80, Federal Republic of Germany
7. Isocommerz GmbH, Kontor Berlin, DDR-115, Berlin-Buch, Lindenberger, German Democratic Republic
8. Medimpex Hungarian Trading Company, Budapest, Hungary
9. N. V. Phillips Duphar Cyclotron and Isotope Laboratories, Petten, The Netherlands

USSR

Techsnabexport Moskva G. 200, USSR, Moskva

Japan

1. Dainabot Radioisotope Laboratory Ltd., 2—7, Nihoinbashi-Honcho, Chuo-Ku, Tokyo
2. Daiichi Radioisotope Laboratories Ltd., Daiichi Seiyaku Bld. No. 1, Chome, Edobashi, Nihein bashi, Chuo-Ku, Tokyo

India

Bhabha Atomic Research Centre, Trombay, Bombay — 400 085

Table 21 (continued)
IMPORTANT LARGE-SCALE MANUFACTURERS OF REACTOR-PRODUCED RADIONUCLIDES AND THEIR PREPARATIONS

Australia

Australian Atomic Energy Commission Research Establishment, Lucas Heights, Australia

[a] Supplies mainly radiopharmaceuticals.
[b] Supplies mainly radioimmunoassay products including labeled compounds of ^{125}I.

including ^{123}I, ^{67}Ga, etc. Further, the commercial production and sale of compact cyclotrons and the development of advanced instrumentation for the in vivo applications of positron emitters have led to a large-scale revival of interest in recent years in the cyclotron for production of isotopes. From current indications and trends in the applications of isotopic tracers in the biomedical sciences, it is apparent that future developments in the field of reactor-produced radionuclides will be mainly confined to improvements in production process and enhancement of specific activity and radionuclidic purity for important isotopes such as ^{125}I, ^{99}Mo, ^{75}Se, etc. Iodine-125 with high purity and isotopic abundance exceeding 95% has recently become available from two or three sources and this has greatly facilitated the development of several iodine-125 tagged steroidal hormones and drugs for RIA. Earlier, these assays were based on the tritium labeled hormones having much lower specific activities. Another important recent development has been the large-scale production of high purity fission product ^{99}Mo for use in the manufacture of ^{99}Mo—^{99m}Tc generators for applications in nuclear medicine. ^{99m}Tc generators based on fission product ^{99}Mo yield pure ^{99m}Tc in the form of sodium pertechnetate having a very high radioactive concentration (exceeding 1 Ci/mℓ) which is a great boon for "bolus" administration in dynamic studies. The earlier method of production based on (n, γ) activation of molybdenum targets had limitations on the specific activity and the quantity of ^{99}Mo that can be loaded on a generator. The availability of ^{75}Se of very high specific activity has opened possibilities of ^{75}Se labeling of many biological products for radioassays.[42] Table 22 indicates some of the important developments in the tracer applications of reactor-produced radionuclides after 1946.

XI. CONCLUSION

Tracer studies with reactor-produced radionuclides have led to spectacular advances in biology and medicine during the last three decades. Many of these radionuclides are now regularly produced on an industrial scale and are made available as pure radiochemicals, labeled compounds, and radiopharmaceuticals. Large-scale manufacturers of these products in many advanced countries have established streamlined production facilities and laid down standards and specifications for these products, thus ensuring a ready supply of all important reactor-produced radionuclides with the requisite purity criteria for various applications. Progress in instrumentation, particularly in nuclear medicine and radioimmunoassays, have also facilitated the widespread applications, both in vivo and in vitro, of radiotracers in the biomedical sciences. Further improvements in production methods for many of these isotopes and expansion in the range and extent of their applications can be expected in the future.

Table 22
IMPORTANT DEVELOPMENTS CONNECTED WITH TRACER APPLICATIONS OF REACTOR-PRODUCED ISOTOPES AFTER 1946

Year	Development	Ref.
July 1951	First Radioisotope Conference held at Oxford. This resulted in considerable dissemination of knowledge of tracer applications.	Sponsored by Atomic Energy Research Establishment, Harwell, UK.
August 1951	Issue of a five year summary of distribution (of isotopes) with bibliography. This resulted in considerable dissemination of knowledge of tracer applications.	United States Atomic Energy Commission, "Isotopes", five year summary of distribution with bibliography.
1951	Development of the rectilinear scanner.	Cassen, B., Curtis, L., Reed, C., and Libby, R., *Nucleonics*, 9, 46, 1951.
1952	Development of the scanner.	Mayneord, W. V. and Newberry, S. P., *Br. J. Radiol.*, 25, 589, 1952.
1954	Second Radioisotope Conference at Oxford. This resulted in considerable dissemination of knowledge of tracer applications.	Sponsored by Atomic Energy Research Establishment, Harwell, UK.
1954	Development of the first "Radioisotope Generator" for ^{132}I.	Stang, L. G., Jr., Tucker, W. D., Doering, R. F., Weiss, A. J., Greene, M. W., and Banks, H. O., *Proceedings of the 1st UNESCO International Conf. Paris 1957*, Pergamon Press, London, 1958, 50.
1955	First International Conference on the Peaceful Applications of Atomic Energy. This resulted in considerable dissemination of knowledge of tracer applications.	Sponsored by United Nations.
1956	Development of the $^{99}\Sigma$Tc generator.	Brookhaven National Laboratory, USA.
1956	Whole body counter 4π liquid scintillator developed (Los Alamos)	Anderson, E. C., *Br. J. Radiol. Suppl.*, 7, 27, 1956.
1956	NaI scintillator whole body counter developed	Marinelli, L. D., *Br. J. Radiol. Suppl.*, 7, 38, 1956.
1956	Development of gamma camera.	Anger, H. O., Mortimer, R. K., and Tobias, C. A., Proceedings of the International Conf. on Peaceful uses of Atomic Energy, 14, 204, 1956.
1958	Second International Conference on the peaceful applications of Atomic Energy. (This resulted in considerable dissemination of knowledge of tracer applications.)	Sponsored by United Nations.
1958	Development of Plastic Scintillator whole-body counter	Bird, P. M. and Burch, P. R. J., *Phys. Med. Biol.*, 2, 217, 1958.
1959	Development of the technique of radioimmunoassay.	Berson, S. A. and Yalow, R. S., *J. Clin. Invest.*, 38, 1996, 1959.

Table 22 (continued)
IMPORTANT DEVELOPMENTS CONNECTED WITH TRACER APPLICATIONS OF REACTOR-PRODUCED ISOTOPES AFTER 1946

Year	Development	Ref.
1962	Radioiodine labeling of hormones by chloramine T method.	Hunter, W. M. and Greenwood, F. C., *Nature (London)*, 194, 495, 1962.
1962	Development of a double antibody separation method in RIA.	Utiger, R. D., Parkar, M. L., and Doughaday, W. H., *J. Clin. Invest.*, 41, 254, 1962.
1964	Third International Conference on the Peaceful applications of Atomic Energy. This resulted in considerable dissemination of knowledge of tracer applications.	Sponsored by United Nations
1966	Development of solid phase RIA.	Calt, K., Niall, H. O., and Tregear, G. N., *Biochem. J.*, 100, 31C, 1966.
1966	Development of the ^{113}Sn—$^{113}\Sigma$In generator.	Gillette, Review of Radioisotope Progress, ORNL Rep. No. 4013, U.S. AEC. Oak Ridge, Tn., 1966.
1967	Automation of RIA.	Bagshawe, K. W., Harris, F. W., and Orr, A. M. Technicon Symposium, Vol. 2, 1967, 53.
1967	Development of immunoradiometric assay.	Wide, L. and Bennich, H., *Lancet*, 2, 1105, 1967.
1968	RIA of haptens.	Oliver, G. C. and Parkar, B. M., *J. Clin. Invest.*, 47, 1035, 1968.
1971	Development of $^{99}\Sigma$Tc labeled polyphosphate for bone scanning.	Subramanian, G. and McAfee, J. G., *Radiology*, 99, 192, 1971.
1971	Fourth International Conference on the Peaceful Uses of Atomic Energy.	Sponsored by United Nations and IAEA
1972	Development of the Bolton-Hunter reagent for labeling in RIA.	Bolton, A. E. and Hunter, W. M., *J. Endocr.*, 55, 30, 1972; *Biochem. J.*, 133, 529, 1973.
1975	Development of in vitro production method for specific antibodies.	Kohler, G. and Milstein, C., *Nature*, 256, 495, 1976; *Eur. J. Immunol.*, 6, 511, 1976.

XII.
PHYSICAL DATA FOR IMPORTANT REACTOR-PRODUCED RADIONUCLIDES OF INTEREST AS TRACERS IN BIOLOGY AND MEDICINE

Radionuclide	Half-life	Mode of decay		Energy of radiations emitted	
^{3}H	12.33 Y	β^- (100%)	β^- 0.018619 MeV (100%)		
^{14}C	5730 Y	β^- (100%)	β^- 0.155 MeV (100%)		
^{18}F	109.8 M	β^+ (96.9%)	β^+ 0.635 MeV (96.9%)		
		EC (3.1%)			
^{24}Na	15.03 h	β^- (100%)	β^- 1.3892 MeV (100%)	γ	1.36853 MeV (100%)
					2.7541 MeV (100%)
					3.8672 MeV (0.061%)
					4.2389 MeV (8.4 × 10^{-4}%)
^{28}Mg	20.93 h	β^- (100%)	β^- 0.459 MeV	γ	0.031 MeV (95%) 0.9722 MeV (<0.19%)
					0.2472MeV(0.053%) 1.3422MeV(56.8%)
					0.4006MeV(37.8%) 1.3728 MeV (4.9%)
					0.6481 MeV (0.089%) 1.5894 MeV(4.9%)
					0.9417 MeV (37.8%) 1.62 MeV (0.3%)
^{32}P	14.28 d	β^- (100%)	β^- 1.711 MeV (100%)		
^{35}S	87.4 d	β^-(100%)	β^- 0.1674 MeV (100%)		
^{42}K	12.36 h	β^-	β^- 3.56 MeV (81.2%)	γ	0.3126 MeV (0.349%)
			1.97 MeV (18.3%)		0.8997 MeV (0.0536%)
					1.525 MeV (18.8%)
^{45}Ca	165 d	β^- (100%)	β^- 0.258 MeV (100%)	γ	1.297 MeV (77%)
^{47}Ca	4.536 d	β^-	β^- 1.9806 MeV (16%)	γ	0.4889 MeV (6.815%)
			1.241 MeV (1%)		0.8079 MeV (6.77%)
			0.684 MeV (83.9%)		
^{51}Cr	27.70 d	EC (100%)		γ	0.32 MeV (10.2%)
^{58}Co	70.8 d	β^+ (14.5%)	β^+ 0.474 MeV	γ	0.811 MeV (99.44%)
		EC (85.5%)			0.86374 MeV (0.686%)
					1.67473 MeV (0.522%)
^{59}Fe	44.56 d	β^-	β^- 1.573 MeV (0.30%)	γ	0.143 MeV (1.02%)
			0.475 MeV (51.2%)		0.192 MeV (3.08%)
			0.273 MeV (48.5%)		0.3348 MeV (0.27%)
					0.3820 MeV (0.018%)
					1.099 MeV (56.5%)

XII.
PHYSICAL DATA FOR IMPORTANT REACTOR-PRODUCED RADIONUCLIDES OF INTEREST AS TRACERS IN BIOLOGY AND MEDICINE (continued)

Radionuclide	Half-life	Mode of decay	Energy of radiations emitted	
^{82}Br	35.34 h	β^-	0.444 MeV	
				1.292 MeV (43.2%)
				1.4817 MeV (0.059%)
		γ		0.09219 Mev (0.748%)
				0.1374 MeV (0.1526%)
				0.221411 Mev (2.27%)
				0.273419 MeV (0.843%)
			0.3329 MeV (0.09%)	
		γ	0.40112 MeV (0.0909%)	1.65029 MeV (0.789%)
			0.554322 MeV (70.72%)	
			0.606317 MeV (1.209%)	
			0.619054 MeV (43.03%)	
			0.698320 MeV (28.606%)	
			0.776489 MeV (83.4%)	
			0.827812 MeV (23.94%)	
			0.95212 MeV (0.37%)	
			1.1801 MeV 1.31%)	
			1.317473 MeV (27.36%)	
			1.47482 MeV (16.6%)	
^{85}Sr	64.8 d	EC		γ 0.514 MeV (99.27%)
^{86}Rb	18.8 d	β^-	1.770 MeV (88%)	γ 1.077 MeV (8.79%)
			0.680 MeV (12%)	
^{90}Y	64.1 h	β^-	2.288 MeV	γ
^{91}Y	58.5 d	β^-	1.543 MeV	γ 1.208 MeV (0.3%)
^{99}Mo	66.02 h	β^-	1.214 MeV (84%)	γ 0.04055 MeV (0.8694%)
			0.840 MeV (2%)	0.14050 MeV (with Tc-99m)
			0.450 MeV (14%)	0.18109 MeV (6.287%)
				0.36645 MeV (1.348%)
			γ 0.7394 MeV (12.6%)	
			0.7778 MeV (4.39%)	
			0.8228 MeV (0.139)	
			0.9610 MeV (0.098%)	

^{99m}Tc	6.02 h	IT	γ	0.141 MeV (89%)
				0.14263 MeV
^{113}Sn	115.1 d	EC	γ	0.255060 MeV (1.82%)
				0.392 MeV (64%)
				0.638030 MeV (0.0015%)
				0.64675 MeV (65 × 10^{-6}%)
^{113m}In	99.47 m	IT	γ	0.392 MeV (64%)
^{125}I	60.25 d	EC	γ	0.035 MeV (6.7%)
^{131}I	8.04 d	β$^-$	β$^-$	0.81 MeV (0.6%)
				0.6065 MeV (86%)
				0.336 MeV (13%)
			γ	0.080183 (2.60%)
				0.17721 (0.264%)
				0.284 (6.04%)
				0.3258 (0.2503%)
				0.6369 (7.24%0
				0.7228 (1.79%)

REFERENCES

1. **Meinhold, H., Herzberg, B., Kaul, A., and Roedler, H. B.,** Radioactive impurities of nuclide generators and estimation of resulting absorbed dose in man in *Proceedings of a Symposium,* Vol. 1, IAEA and W.H.O., Copenhagen, 1973, 50.
2. **Colombetti, L. G.,** Performance of ^{99m}Tc generating systems, in *Quality control in Nuclear Medicine,* Rhodes, B. A., Ed., The C. V. Mosby Company, St. Louis, 1977, 190.
3. **Mani, R. S. and Gopal, N. G. S.,** Industrial production of radiopharmaceuticals, in *Radiopharmacy,* Tubis, M. and Wolf, W., Eds., John Wiley & Sons, New York, 1976, 508.
4. **Tubis, M.,** Quality control of radiopharmaceuticals, in *Radiopharmacy,* Tubis, M. and Wolf, W., Eds., John Wiley & Sons, New York, 1976, 556.
5. **Subramanian, G.,** Radioisotope generators, in *Radiopharmacy,* Tubis, M., and Wolf, W., Eds, John Wiley & Sons, New York, 1976, 267.
6. **Crosby, E. H.,** Radiochemical purity of short-lived ^{99m}Tc from commercial suppliers, *Radiology,* 93, 435 1969.
7. Radioisotope Production and Quality Control, Technical Reports Series No. 128, IAEA, Vienna, 1971.
8. **Cohen, Y. and Merlin, L.,** Requirements for product quality control, in *Radioisotope Production and Quality Control* Technical Reports Series No. 128, IAEA, Vienna, 1971, 930.
9. **Cifka, J.,** Radiochemical purity and stability of some radiopharmaceuticals, in *Analytical Control of Radiopharmaceuticals,* Proc. Panel, IAEA, Vienna, 1969, 177.
10. **Cohen, Y.,** Purity criteria and general specification of Radiopharmaceuticals, in *Analytical Control of Radiopharmaceuticals,* Proc. Panel, IAEA, Vienna, 1969, 1.
11. **Iya, V. K., Mani, R. S., and Desai, C. N.,** Preparation of labelled molecules, in *Radioisotope Production and Quality Control,* Technical Reports Series No. 128, IAEA, Vienna, 1971, 842.
12. **Roche, J.,** The development of modern biochemistry and the use of radioisotopes, in *Proceedings of a Conference on Methods of Preparing and Storing Marked Molecules,* Euratom, Brussels, 1963, 745.
13. **Bayly, R. J., Evans, and Glover, J. S.,** Synthesis of labelled compounds, in *Radiopharmacy,* Tubis, M. and Wolf, W., Eds., John Wiley & Sons, New York, 1976, 357.
14. **Hunter, W. M.,** Iodination of protein compounds, in *Radioactive Pharmaceuticals,* USAEC Symposium series No. 9, Oak Ridge, Tenn., 1966, 245.
15. **Rosa, U. and Malvano, R.,** Labelled compounds for radioimmunoassay procedures, in *Radiopharmaceuticals and Labelled Compounds,* Vol. 2, IAEA, and W.H.O., Copenhagen, 1973, 94.
16. **Burgess, J. S. and Partington, E. J.,** Radiation Decomposition Effects in Aqueous Solutions of Carrier-Free Sodium Iodide I-131, UKAEA Report RCC/R 98, 1960.
17. *The Radiochemical Manual,* 2nd ed., The Radiochemical Center, Amersham, England, 1966, 98.
18. **Greenwood, F. C., Hunter, W. M., and Glover, J. S.,** The preparation of ^{131}I-labelled human growth hormone of high specific radioactivity, *Biochem. J.,* 89, 114, 1963.
19. **Mani, R. S. and Chowdhary, S. Y.,** Preparation of ^{51}Cr of high specific activity for medical use, *Curr. Sci. (India),* 35, 230, 1966.
20. **Mani, R. S.,** Extraction of high specific activity ^{51}Cr from pile-irradiated K_2CrO_4, *Int. J. Appl. Radiat. Isotopes,* 14, 327, 1963.
21. **Arrol, W. J., Wilson, E. J., Evans, C., Chadwick, J., and Eakins, J.,** The preparation and possible industrial uses of Kr^{85} and tritium, in *Radioisotope Conference 1954 — Proceedings of the 2nd Conference,* Butterworths, London, 1954, 61.
22. **Massey, B. J.,** ORNL Rep. No. 2238, Oak Ridge National Laboratories, Oak Ridge, Tenn., 1957.
23. **Abraham, B. M.,** U.S. Patent 3,100,184 (1963).
24. **Shields, R. P.,** The Production of ^{14}C by the Be_3N_2 Process, ORNL Report, Oak Ridge National Laboratories, Oak Ridge, Tenn., 1962.
25. **Charlton, J. C. and Evans, C. C.,** British Patent 784,125 (1957).
26. **Rupp, A. F.,** *Large-Scale Production of Radioisotopes,* Vol. 14, United Nations, New York, 1956, 74.
27. **Mani, R. S. and Majali, A. B.,** Production of carrier-free ^{32}P, *Ind. J. Chem.,* 4, 391, 1966.
28. Radioisotope Production and Quality Control, Technical Reports Series No. 128, IAEA, Vienna, 1971, 381.
29. **Evans, C. C. and Stevenson, J.,** Production of Radioactive Phosphorus, British Patent 765,489 (1957).
30. Radioisotope Production and Quality Control, Technical Reports Series No. 128, IAEA, Vienna, 1971, 370.
31. **Rupp, A. F.,** *Large-Scale Production of Radioisotopes,* Vol. 14, United Nations, New York, 1956, 78.
32. **Veljkovic, S. R. and Milenkovic, S. M.,** *International Conference on Peaceful Uses of Atomic Energy,* Vol. 20, United Nations, New York, 1958, 45
33. **Mani, R. S., Subramanian, M., and Iya, V. K.,** Production of ^{36}Cl and carrier-free ^{35}S from pile-irradiated KCl, *Ind. J. Technol.,* 1, 1, 24, 1963.

34. **Harper, P. V., Seimens, W. D., Lathrop, K. A., and Endlich, H.,** Production and uses of ^{125}I, *J. Nucl. Med.*, 4, 277, 1963
35. Radioisotope Production and Quality Control, Technical Reports Series No. 128, IAEA, Vienna, 1971, 240.
36. Radioisotope Production and Quality Control, Technical Reports Series No. 128, IAEA, Vienna, 1971, 258.
37. **Evans, C. C. and Stevenson, J.,** Improvements in or Relating to Production of Radioactive Iodine-131, British Patent 763,865 (1956).
38. **Alekseev, R. I. and Polevaya, O. N.,** Separation of ^{99}Mo from a mixture of U fission products, *Radiokhimiya*, 3, 458, 1961.
39. **Constant, R.,** Determination of the specific activity of carrier-free ^{131}I, *Int. J. Appl. Radiat. Isotopes*, 16, 447, 1965.
40. **Arino, H. and Kramer, H. H.,** Determination of specific activity of ^{131}I solution via an iodide electrode, *Nucl. Appl.*, 4, 356, 1968.
41. *ATA Restricted Articles Regulations,* 21st ed., International Air Transport Association, Geneva, 1978, 247.
42. Medical Products 1978/1979, The Radiochemical Center, Amersham, Buckinghamshire, England, 1978, 16: *Radiochemicals January 1981 Catalogue,* The Radiochemical Center, Amersham, Buckinghamshire, England, 1981, 123.
43. Radiation Protection Procedures, Safety Series No. 38, International Atomic Energy Agency, Vienna, 1973, 83.
44. *The United States Pharmacopeia,* 20th ed., Mack Publishing, Easton, Pa., 1980, 764.
45. **Lederer, M. and Shirley, V., Eds.,** *Table of Isotopes,* 7th ed., John Wiley & Sons, New York, 1978.
46. Handbook of Nuclear Activation Cross-Section, Technical Reports Series, IAEA, Vienna, 1974.
47. **Ridder, B. F.,** Compilation of Fission Product Yields, NEDO 12154;3(B), ENDE 292 DRF 267; 0005, Vallecitos Nuclear Center, 1980.
48. **Aliev, A. I., Drynkin, V. I., Leipunskaya, D. I., and Kasatkin, V. A.,** *Handbook of Nuclear Data for Neutron Activation Analysis,* Atomizdate, Moskva, 1969.
49. **Schett, A., Okamoto, K., Lesca, L., and Frohner, F. H.,** Compilation of Threshold Reaction Neutron Cross-Sections for Neutron Dosimetry and Other Applications, OECD/NEA Centre de Compilation de Donnees Neutroniques, CBNM, 1974.
50. **Shikata, E.,** Preparation of ^{18}F; separation on alumina column and chemical form of ^{18}F obtained, *J. Nucl. Sci. Tech.,* 1, 183, 1964.
51. **Mani, R. S. and Narasimhan, D. V. S.,** Development of kits for short-lived generator-produced radioisotopes, in *Radiopharmaceuticals and Labelled Compounds,* IAEA, and W.H.O., Copenhagen, 1973, 141.

Chapter 2

SHORT-LIVED POSITRON EMITTING RADIONUCLIDES

W. Vaalburg and A. M. J. Paans

TABLE OF CONTENTS

I.	Introduction	48
	A. Short-Lived Radionuclides	48
	B. Decay by Positron Emission	48
	C. Application of Short-Lived Positron Emitters	49
	D. Availability and Specific Activity	50
	E. Preparation of Positron Emitting Short-Lived Labeled Compounds	51
	F. Medical Imaging and Quantification	53
	1. Single Section Devices	53
	2. Multisection Devices	54
	G. Advantages and Disadvantages	54
II.	Production of Carbon-11	55
	A. Decay of Carbon-11	55
	B. Choice of Nuclear Production Reactions	55
	C. Production of ^{11}C-Labeled Oxides	57
	1. ^{11}C-Labeled Carbon Dioxide	57
	2. ^{11}C-Labeled Carbon Monoxide	59
	D. ^{11}C-Labeled Cyanide	60
	E. ^{11}C-Labeled Acetylene	62
	F. ^{11}C-Labeled Methyl Iodide	62
	G. ^{11}C-Labeled Methyllithium	63
	H. ^{11}C-Labeled Formaldehyde	64
	I. ^{11}C-Labeled Phosgene	65
III.	Production of Nitrogen-13	65
	A. Decay of Nitrogen-13	65
	B. Choice of Nuclear Production Reactions	65
	C. ^{13}N-Labeled Nitrates and Nitrites	66
	D. ^{13}N-Labeled Ammonia	67
	E. ^{13}N-Labeled Molecular Nitrogen	70
	1. The ^{12}C(d,n)^{13}N Reaction on Solid Targets	70
	2. The ^{12}C(d,n)^{13}N Reaction on Carbon Dioxide	72
	3. The (p,n) Reaction on Enriched Carbon-13	72
	4. The (p,α) Reaction on Water	73
	F. ^{13}N-Labeled Nitrogen Oxides	74
IV.	Production of Oxygen-15 and Oxygen-14	74
	A. Decay of Oxygen-15 and Oxygen-14	74
	B. Choice of Nuclear Production Reactions	74
	C. ^{15}O-Labeled Molecular Oxygen	76
	D. ^{15}O-Labeled Carbon Dioxide	78
	E. ^{15}O-Labeled Carbon Monoxide	78

F. ^{15}O-Labeled Water.. 79
G. ^{14}O-Labeled Molecular Oxygen and ^{14}O-Labeled Water 79

V.   Labeling Organic Compounds with Carbon-11 and Nitrogen-13 80
    A.   The Literature 80
    B.   Carbon-11 Labeled Organic Compounds................................. 81
    C.   Nitrogen-13 Labeled Organic Compounds 83

VI. Generators for Short-Lived Positron Emitters 84
    A.   The ^{52}Fe-$^{52}\Sigma$Mn System 84
    B.   The ^{62}Zn-^{62}Cu System 84
    C.   The ^{68}Ge-^{68}Ga System 84
    D.   The ^{82}Sr-^{82}Rb System 85
    E.   The ^{122}Xe-^{122}I System... 85

VII. Some Other Short-Lived Positron-Emitting Nuclides 85
    A.   Neon-19 85
    B.   Phosphorus-30 86
    C.   Potassium-38 86
    D.   Iron-52 87
    E.   Rubidium-81... 88

References... 89

I. INTRODUCTION

A. Short-Lived Radionuclides

When a nuclear physicist, investigating radionuclides far from the stability line, is asked to give an example of a short-lived radionuclide, he will probably ask you to specify short-lived or he will mention a radionuclide with a half-life in the sub-millisecond region. In his eyes carbon-11, with its 20 min half-life, is a long-lived nuclide. When the same question is asked to a biochemist, used to working with carbon-14, he will probably give carbon-11 as an example of a short-lived nuclide. So it is clear that the term short-lived needs further definition. Some authors involved in biological and biomedical research consider a radionuclide short-lived when its half-life is less than 15 hr. Others define the border line at 12 hr or even shorter. Another approach for definition is the criterium of transportability. With this criterium a radionuclide is defined as short-lived, when because of the half-life, the maximum distance between production facility and user is limited to a few kilometers. From this point of view technetium-99m is not short-lived, because it can be transported as generator system all over the world. Anyway, the aim of this criterium of transportability is to stress that short-lived nuclides can only be used after on-site production. We choose to define a radionuclide as short-lived when its half-life is shorter than 15 hr.

B. Decay by Positron Emission

In the beta decay process a positive or negative charged electron is emitted from the nucleus. In all cases a neutrino is simultaneously emitted. The electron and the positron, a positively charged electron, are antiparticles and they can annihilate when brought in close contact. All neutron-rich nuclei will decay via β^- emission along its isobar to a stable isotope. All proton-rich nuclei will decay with the emission of β^+ particles or by electron capture. The two basic processes in the decay of proton-rich nuclei are

$$p \longrightarrow n + \beta^+ + \nu \qquad \beta^+ \text{ decay}$$

$$p + e^- \longrightarrow n + \nu \qquad \text{electron capture (EC)}$$

The symbol β^+ is used for the positron created in the decay process and the symbol e^- is used for an orbital electron. The symbol ν represents a neutrino. The neutrino, first proposed by Pauli in 1931, is a neutral particle with zero rest mass.[1] The kinetic energy available for the decay product can be calculated from the atomic rest energies. In the decay process by positron emission the nuclear charge is decreased by one unit. The resulting ion therefore has a negative charge of one unit. The maximum energy available for the positron ($Q_{\beta}+$) is given by the difference in the rest energies of both atoms, before and after the decay, and is given by:

$$Q_{\beta^+} = M(A, Z + 1) c^2 - M(A, Z) c^2 - 2m_o c^2 + I \qquad (1)$$

$M(A, Z)$ is the mass of an atom with mass number A and atomic number Z. I is the binding energy of the last atomic electron and is so small that it is usually neglected. The rest mass of an electron or a positron is denoted by m_o. From Equation 1 it is obvious that an energy excess of $2m_o c^2 = 1022$ keV has to be available in order to have positron emission. This minimum energy of 1022 keV required, is the main reason why ^{123}I is decaying by electron capture for virtually 100% ($I_{\beta}+ < 0.01\%$).[2,3] The difference in the mass excess between ^{123}I and ^{123}Te is 1200 keV. After emission of the positron it will be slowed down in matter and eventually pair with one of the electrons within its range. Before the two will annihilate the system exists as a positronium atom. This is a configuration like a hydrogen atom in which the proton nucleus is replaced by a positron. The ground state configuration of the positronium can be a singlet or a triplet. Since the selection rules operate different on both states the annihilation process is also different. The mean life time of the singlet state is 8 ns and decays into two quanta with each an energy of $m_o c^2$ (511 keV).[4] The triplet state has a mean life time of 7 µs and decays into three quanta. The sum of the energies of the three quanta is $2m_o c^2$ (1022 keV). In about 0.3% of the decays a positronium in the triplet state is formed,[5] so virtually all positrons decay from the singlet state resulting in two quanta of 511 keV each. The ratio between the two-quanta and three-quanta decay of the positronium can be influenced by the chemical composition of the matter in which the formation and the decay of the positronium takes place.[6] If the momentum of the center of mass of the two-body system is zero in the laboratory system the two quanta will appear exactly collinear. Because most annihilations will take place at nonzero momentum a finite width, in the order of 0.5° will be present in the angular distribution about the mean angle of 180° between the two annihilation quanta.[7]

C. Application of Short-Lived Positron Emitters

Radionuclides are applied in many different areas including medicine. In medicine they are used for therapy as well as for diagnosis. In therapy the destructive effect of radiation of the radionuclides administered to the patient is used. For diagnosis, however, information must be obtained as accurately as is clinically necessary and this destructive effect minimized. If for medical application a radioactive compound is used in vivo, it is called a radiopharmaceutical, which is defined by its chemical and by its radioactive properties. The chemical structure determines the mode of localization of the material in the patient, or the way it is involved in metabolism. The type of radiation emitted by the radiopharmaceutical is dependent upon the nuclear properties of the radioactive label. For in vivo measurement it is necessary that the radiation can be detected outside the body of the patient; the radiation must be externally detectable. Many diagnostic in vivo methods are based on scintigraphic techniques. These are techniques by which an image of the tissue distribution of the radioactivity administered to the patient is obtained. When positron emitting radionuclides in combination with computerized tomographic systems are used, the distribution of the radioactivity can be obtained at different levels of the body and the distribution can be quantitated.

The amount of radioactivity remaining in the body, after the diagnostic information is obtained, delivers an unnecessary contribution to the total absorbed radiation dose so the radiation must disappear as soon as possible. The disappearance occurs by biological clearance of the radioactive compounds, and by decay. The result is called the effective half-life. Reduction of the effective half-life can be achieved by selecting the proper chemical form of the radiopharmaceutical and by labeling the compound with a short-lived radionuclide. Besides the reduction of radiation dose a short effective half-life opens the possibility of repeated investigations under different physiological conditions, or the use of several radiopharmaceuticals in sequence. Another approach for diagnosis is to measure quantitatively the function of an organ or a metabolic process by following the fate of the radioactive compound in the organism. In this context positron emitters give new dimensions to nuclear medicine and pharmacology. For probing metabolism the most evident approach is to use labeled metabolites as radiopharmaceuticals: therefore, the use of labeled compounds with exactly the same chemical identity as the compound involved in metabolism has distinct advantages. From deviations of the normal pattern diagnostic information can be obtained.

Another approach is the use of labeled analogues which are sufficiently close in chemical structure to the metabolite that the organism, to a certain level, cannot discriminate between the natural compound and the labeled analogue. Also labeled competitive antimetabolites can be used. These compounds block a certain step in metabolism by competitive action which results in accumulation in a certain stage. In this way radioactive antimetabolites can be used to isolate certain steps in metabolism.

Sometimes fruitful results can be obtained with labeled antagonists. Their success as radiopharmaceutical depends upon the mode of action. If, for instance, the antagonist competes for specific receptors, then the labeled version has potential to trace these receptors. It is clear that when more different radionuclides are available, more radioactive compounds can be prepared. Short-lived positron emitters play in this context an unique role.

Many interesting compounds mentioned above do not contain elements other than carbon, hydrogen, oxygen, and nitrogen. The chart of nuclides, however, shows that of these elements only a limited number of isotopes with externally detectable radiation and useful half-lives are known: carbon-11, nitrogen-13, oxygen-14, and oxygen-15. They all emit positrons and are short-lived. Besides more complex organic compounds simple labeled inorganic molecules like CO, CO_2, N_2, and NH_3 are used extensively.[8,9] The application of these radionuclides open new avenues, not only for medical purposes. They are also powerful tools for in vivo pharmacokinetic studies and in biology. Nitrogen-13 for instance is used in nitrogen fixation studies and in soil denitrification measurements. These and many other applications make short-lived positron emitters intriguing.

D. Availability and Specific Activity

A convenient way to make these radionuclides available at long distances from the production site is by a nuclide generator system ("radioactive cow"). The principle of any nuclide generator system is a mother-daughter relationship, whereby the short-lived daughter nuclide is separated from its longer-lived mother nuclide at the appropriate time. However, the number of possible generator systems is limited. A different approach in making radionuclides of short half-life available is to bring the source of production to the user. In general, positron emitters are produced by charged particle reactions. For the acceleration of positively charged particles cyclotrons, Van de Graaff generators or linear accelerators are used. Cyclotrons are preferred to Van de Graaff generators because of the higher maximum particle energy and the high beam currents achievable.

Nuclear reactions induced with charged particles very often result in product nuclides of a different element than the target nuclide. This implies that theoretically carrier-free products can be made. Carrier-freeness is very often to be preferred in biomedical research because radioactive compounds labeled with these nuclides may be added to biological systems without significantly changing the physiology or biochemistry of the stable element or compound already present. Therefore, even very toxic compounds can be applied. However, in practice true carrier-freeness is difficult to achieve, because the risk of contamination with stable nuclides during production of the radioactivity and the preparation of the labeled compound is hard to avoid. This problem is not always realized. When a preparation of a carrier-free compound is reported very often the author is using this term in the sense that no carrier is added during the whole procedure. Very often the product is not really carrier-free. Therefore, to avoid confusion, it is proposed to use the term carrier-free (CF) when it is experimentally verified that the product is carrier-free in the sense that no carrier is present at all. In all other cases the terminology has to be No Carrier Added (NCA) or Carrier Added (CA).[10]

E. Preparation of Positron Emitting Short-Lived Labeled Compounds

Compounds labeled with short-lived positron emitting nuclides include organic as well as inorganic molecules. In this part only some general aspects of labeling organic compounds will be discussed. Inorganic radioactive compounds like $^{11}CO_2, ^{11}CO_2, H_2^{15}O$, and other simple molecules, are very often produced directly in the target or on line. Syntheses of ^{15}O-labeled organic compounds, as far as we know, are not reported until now. The general approach for labeling is a multistep process:

- The direct in-target production of a radioactive synthetic precursor
- The synthesis of the labeled compound from this precursor
- The purification and, if necessary, sterilization of the product

It is obvious that because of the half-life of the radionuclides involved, for all these steps the time-factor has to be considered carefully. A rule of thumb is that the compound must be available for application within three half-lives after the radionuclide production is finished. The moment the production of radioactivity is stopped, is mostly referred as the end of bombardment (EOB). Very often the time scale of a preparation is related to EOB. When the compounds are prepared for application in nuclear medicine or in vivo radiopharmacology the products must be of radiopharmaceutical quality and very high specific activity. As mentioned already the radioactive synthetic starting material can be prepared either directly in the target or by an on-line method from products formed in the target. For example $^{11}CO_2$ is formed directly while $^{11}CH_3I$ can be prepared on line from $^{11}CO_2$.

The product spectrum in the target is a result of primary product formation by hot atom reactions and radiolysis. When a nuclear reaction is induced in a target, the energy released is carried away by the emission of particles or radiation and by the newly formed atom. The product nucleus is called a "hot atom" because it is in an unusually high energy state. The kinetic energy part is called recoil energy. The recoil energy can vary from 0.1 eV or less to several MeV. The energy of the formed atom is in most cases sufficiently high to break the bond with the mother molecule. The hot atom moves into the surrounding material. The increased energy of the atom is often manifest in the form of electronic excitation, ionization and a high kinetic energy. Through electronic interactions and scattering phenomena the atom is slowed down and reaches an energy state, where elastic and inelastic collisions with molecules can occur. Before the atom is in thermal equilibrium with its surrounding, it reacts chemically with the surrounding material to give one or more radioactive products. This process is called a hot-atom reaction or a recoil reaction. Because the energy of the atom, at which a hot atom reaction takes place, is much higher than the energy of the thermal atom, it undergoes a reaction that will be different from the reaction of a thermal atom. Some criteria are applied to set a recoil reaction apart from a thermal reaction. A recoil reaction is

- Temperature insensitive
- Phase independent
- Dependent on radical scavengers
- Dependent on moderators (e.g., inert gases).

General information about hot atom chemistry can be found in a review by Wolf[11] and in a monograph by Stöcklin.[12]

To induce a nuclear reaction a large amount of radiation is involved. It is unavoidable that the target material is damaged by this radiation or by the recoiling atom: radiolytic decomposition. Impurities in the bombarded material play in this case an important role during irradiation. Impurities, for example, can be excited and can react with the target material or with recoiling atoms. Certainly they affect the spectrum of compounds obtained after irradiation, particularly when the target compound is bombarded in the gas phase. By selecting the proper target material, nuclear reaction and irradiation conditions the chemical form of the radioactive products can be controlled to some extent. Preparations of a labeled complex organic compound by direct bombardment of the nonradioactive analogue is not very practical because a highly impure product with a low specific activity will be obtained.

In planning a synthesis several factors have to be considered:

- Place of the radioactive label in the molecule
- Specific activity
- Amount of radioactive product desired

The synthesis itself can be either inorganic, organic, biochemical, or electrochemical. Some aspects of the methods used are similar to those for the preparation of carbon-14 compounds. However, the preparation of carrier-free compounds, including the problems associated with carrier contamination, are completely different. Other differences are related with the half-life of the radionuclides and the limited number of radioactive precursors with which the preparation can be started. Moreover the precursor must be produced immediately before use. Another challenging feature is the 511 keV gamma ray radiation as consequence of the positron emission; all manipulations must be carried out automatically or under remote control behind heavy lead shielding or in hot cells. Because of the fast decay of the radioactivity the chemical reaction has to be selected not only on the chemical yield but also on the time required to reach its yield. It is possible that a fast reaction with a low chemical yield gives a higher radiochemical yield than a slow reaction with a high chemical yield.[13] Biochemical methods have the advantage over the other methods mentioned that complex chemical compounds can be obtained very often in the desired stereochemical form. This may be essential for biomedical application. Specific activity as well as apyrogenicity may be a problem when biochemical methods are used. On the other hand, methods such as those which make use of immobilized enzymes, make automatization easier. This has a positive influence on the reduction of the radiation dose received by the chemist. The main techniques used for purification and analysis are gas and high pressure liquid chromatography. Besides a nondestructive mass detector the apparatus must be equipped with a radioactivity detector.[14]

When the labeled compound is prepared for in vivo use in human or in test animals, the final product must be of radiopharmaceutical quality. It must be sterile, isotonic, and of an appropriate pH and may not contain pyrogens. Because of the half-life, analysis prior to use is not possible. Retrospective anaysis is acceptable. Therefore, sterile and pyrogen free chemicals and equipment have to be used. Risks to introduce pyrogens during the preparation have to be avoided rigorously. The end-product can be sterilized by ultra membrane filtration. When after several test runs it is shown by analysis that an injectable product is obtained, the same protocol has to be used for the preparation of the radiopharmaceutical.

F. Medical Imaging and Quantification

The conventional imaging device in nuclear medicine is the scintillation camera. Interaction of radiation with the NaI(Tl) crystal results in a flash of light and is seen by the photomultiplier tubes mounted on the crystal. The position information of the actual place of interaction is obtained via a resistor or a delay-line network on the outputs of the photomultiplier tubes. A collimator, placed in front of the crystal, is used to select the photons. A parallel hole collimator for example allows for a one to one image of the object on the crystal. The resolution and the efficiency of a collimator is determined by the size, the number, and the length of the holes and by the septa thickness between the holes. A parallel hole collimator used for imaging of $^{99}\Sigma$Tc-labeled radiopharmaceuticals has an efficiency of approximately 0.01%. If the gamma ray energy of the radionuclides used increases, the septa thickness has to be increased too in order to avoid cross-talk between different holes. By increasing the gamma ray energy from 140 keV to 511 keV the half-value thickness increases from nearly 0.3 mm to almost 4 mm. A lead collimator for the imaging of high energy gamma rays, as for instance the annihilation radiation of positron emitters, will have much less holes, due to the increased septa thickness, resulting in a lower efficiency and/or resolution. Using a tungsten instead of a lead collimator results in an increase of efficiency due to the thinner septa, but the quality of the images obtained is still not acceptable as is shown by Walsh.[15]

Another way of imaging positron emitters is to employ the fact that annihilation radiation in virtually all cases is a two quanta annihilation. By using no collimator at all, but more detectors and coincidence electronics, an inherently tomographic imaging system is obtained. When in two detectors an event is detected at the same moment we know that the annihilation took place on a straight line between these two detectors. If the radionuclides are concentrated in only one spot in the body all the lines connecting the detected coincident events will pass through this one spot. On this principle positron emission tomography is based. Positron imaging devices designed up till now can be divided into single section devices and multiple section devices.

1. Single Section Devices

A single section device consists of a ring or a hexagon of detectors, imaging a transverse section of the object. Rankowitz and co-workers built in Brookhaven[16] the first single section device consisting of 32 crystals. Two devices have been built in Montreal. The first with 32 NaI (Tl) crystals[17] and the second with 64 bismuth germanate (BGO) detectors. The Positron Emission Transaxial Tomograph (PETT) was developed by Ter-Pogossian and co-workers at Washington University.[18,19] The detectors are built in a hexagon which is translated and rotated during the data acquisition. Based on this device Ortec designed the ECAT system as described by Phelps.[20] Cho at UCLA developed a circular system with 64 NaI (Tl) detectors[21] with a diameter of 47 cm. In Sweden a ring system of 92 detectors has been designed.[22] The system wobbles around a small center circle in order to achieve a more dense linear sampling. Derenzo at the Donner Laboratory in Berkeley constructed a ring system with 280 NaI (Tl) detectors.[23] The spatial resolution of the positron imaging devices mentioned above varies from 10 mm to 18 mm FWHM while the sensitivity, of course depending on the diameter of the device, is in the order of 20 Hz/μCi.

2. Multisection Devices

Brownell and co-workers developed a positron imaging system consisting of two detector banks, the Massachusetts General Hospital positron camera, which can rotate around the object.[24,25] Each detector bank contains an array of 12 × 12 NaI (Tl) detectors and 82 photomultiplier tubes.[26] Each detector bank is able to handle a single count rate of 10 MHz. Longitudinal tomograms can be generated by simple back projection. By rotating the system around the object multiple transverse sections can be obtained.[27] Muehllehner[28,29] developed a dual head scintillation camera, as suggested by Anger[30] in 1959. The system consists of two large field of view scintillation cameras in rotatable gantry. To increase the efficiency 2.54 cm thick crystals were used and the electronics were modified in order to allow a higher count rate. The efficiency of this system is 500 Hz/μCi, due to the large solid angle subtended. In a stationary position only longitudinal tomograms can be produced while by rotating the system around the object multiple transverse sections can be obtained. The spatial resolution of 10 mm FWHM is mainly due to the use of 2.54 cm thick crystals. When standard crystals of 1.27 cm are used a spatial resolution of 5.5 mm FWHM can be obtained,[31] however, at the cost of a lower efficiency (200 Hz/μCi). Multiwire chambers, a common detector in the field of high energy physics, are used as a positron imaging device. Since the efficiency of these gas-filled proportional chambers for 511 keV gamma radiation is virtually zero the annihilation radiation is converted into electrons by lead converters. The electrons are then drifted into the chamber and detected. Perez-Mendez and co-workers[32] achieved a resolution of 6 mm FWHM with such a system but with a much lower efficiency than with a system as developed by Brownell or Muehllehner. Also the resolving time of this multiwire system is rather long (300 ns). Jeavons[33] and co-workers developed a positron camera with a spatial resolution of 2 mm FWHM. With this spatial resolution the energy of the positron can be a limitation for the resolution. The intrinsic resolution of the system is mainly determined by the lead converters. The resolving time of 100 nsec can be influenced by the gas mixtures used and the drift velocity of the electrons. A resolving time of 20 ns and an inherently reduction of the accidental coincident events, seems to be obtainable.[34] Ter-Pogossian and his co-workers developed a multislice version of the PETT III device, the PETT IV.[35] Multiple sections can be imaged by the use of long crystals viewed at both ends by a photomultiplier tube. Also a device, PETT V, with a better spatial resolution (7.5 mm FWHM) for brain imaging is under construction.[36]

In systems with a 2π geometry, like ring systems, quantification is possible if a correction for the attenuation of the radiation is applied. The attenuation correction can be measured via a transmission image of the object. This possibility of quantification and the use of positron-emitting radiopharmaceuticals opens a new field of quantitative metabolic dynamic function studies. In stationary systems with two detectors[28,29,32-34] this quantification is not possible due to the limited angle. The images obtained via simple back projection will show a blurring due to out of focus radioactivity. Using back projection and a Fourier deconvolution process[37,38] this blurring can be removed. In the text above the positron imaging devices are mentioned briefly. An elaborate discussion on positron emission tomography is given by Budinger.[39]

G. Advantages and Disadvantages

The biomedical application of short-lived positron-emitting radionuclides has several advantages compared to the application of more conventional radionuclides. The positron emission is detectable outside the body barrier by means of the two 511 keV photons resulting from the annihilation of the positron. This decay process in combination with computerized positron-emission tomography makes noninvasive three dimensional in vivo measurements possible. The most intriguing advantage is perhaps that this tomographic information can be obtained quantitatively.

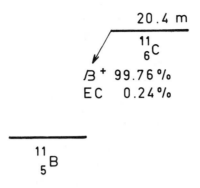

FIGURE 1. Decay scheme of carbon-11.

Other advantages are that the short half-life results in low radiation doses for the patient and the option of repeating the diagnostic procedure within short time intervals. If necessary higher doses of radioactivity can be administered within the safety limits. This can result in more precise information. Some elements, important for biomedical research and investigations only have short-lived positron-emitting radionuclides. The availability of nuclides like carbon-11, nitrogen-13, and oxygen-15 essentially increases the range of valuable radiopharmaceuticals because metabolic organic compounds can be prepared with the same chemical and biochemical properties as the nonradioactive analogues. In principle these compounds can be prepared carrier-free. Because of this possibility extreme toxic compounds can be considered for human use. Using these radionuclides environmental problems such as waste disposal are minimal.

As disadvantage may be mentioned that the radioactivity must be prepared immediately before application, which makes the access to an in-house charged particle accelerator like a medical or compact cyclotron, necessary. It is obvious that application of these nuclides is not possible if information must be collected over a long period of time compared to the half-life of the nuclide. For the preparation of useful amounts of labeled product large amounts of radioactivity have to be handled with as consequence that for the protection of the personnel, automation and remote control of the procedures are necessary. Also heavy lead shielding of 5 to 10 cm has to be installed.

II. PRODUCTION OF CARBON-11

A. Decay of Carbon-11

Carbon-11 is the longest-lived positron-emitting isotope of carbon. This radionuclide decays with a half-life of 20.4 m to stable boron-11.[40] The decay scheme is given in Figure 1. In 99.76% of the cases the decay is by positron emission and only in 0.24% by electron capture. The maximum positron energy is 0.96 MeV and the corresponding range of positrons of this energy in water is 4 mm.

B. Choice of Nuclear Production Reactions

The relevant reactions for the production of carbon-11 are summarized in Table 1. In this table both theoretical and experimental values, if available, of the cross-sections are given. Theoretical values are calculated according to the statistical model as described by Weisskopf and Ewing.[41] The actual calculations were performed with the code ALICE.[42] The cross-sections for the $^{12}C(\gamma,n)^{11}C$ reaction have been measured up to 1 GeV. The highest cross-section of 4mb/eq was reached at 1 GeV.[43] The cross-section is given in mbarn per equivalent quantum (mb/eq), a unit in which is accounted for the bremsstrahlung end-point energy, the bremsstrahlung cross-section, and the cross-section of the reaction. Of the proton-induced

Table 1
RELEVANT REACTIONS FOR THE PRODUCTION OF ^{11}C

Particle	Reaction	Q-value (MeV)	E(thresh.) (MeV)	σ(th.)[a] (mb)	σ(exp) (mb)	Ref.
γ	^{12}C(γ,n)^{11}C	−18.7	18.7		4[b]	43
p	^{11}B(p,n)^{11}C	−2.8	3.0	150	100	45
					350	46
	^{12}C(p,pn)^{11}C	−18.7	20.3	100	100	47—51
	^{14}N(p,α)^{11}C	−2.9	3.1	150	250	52
					180	53
d	^{10}B(d,n)^{11}C	6.5	0	200	180	54
					260	55
	^{11}B(d,2n)^{11}C	−5.0	5.9	60	48	54
	^{12}C(d,p2n)^{11}C	−20.9	24.4	50	61	56,57
^{3}He	^{9}Be(^{3}He,n)^{11}C	7.6	0	190	6	58
					113	59
	^{10}B(^{3}He,pn)^{11}C	1.2	0	210	285	58
	^{11}B(^{3}He,p2n)^{11}C	−1.8	2.3	35		
	^{12}C(^{3}He,^{4}He)^{11}C	1.9	0	240	260	58
					365	59
	^{16}O(^{3}He,2^{4}He)^{11}C	−5.3	6.3	70	49	58
^{4}He	^{9}Be(^{4}He,2n)^{11}C	−13.0	18.8	55	17	54
	^{10}B(^{4}He,p2n)^{11}C	−19.6	27.4	50		
	^{11}B(^{4}He,p3n)^{11}C	−31.1	42.4	17		
	^{12}C(^{4}He,^{4}He, n)^{11}C	−18.7	24.9	110	48	60

[a] Cross-section calculated according to the statistical model.
[b] Cross-section is given in mb per equivalent quantum.

reactions the reaction on boron-11 has the advantage of a low threshold energy and a high cross-section. If natural target material is used, abundance of boron-11 is 80%, also carbon-10 with a half-life of 19.3 s will be produced. The required more complex target technology may be a disadvantage.[44] The ^{12}C(p,pn)^{11}C reaction has the disadvantage of a high threshold energy and the fact that only a low specific activity is obtainable. The proton-induced reaction on nitrogen-14 has the advantage of a low threshold energy, a high cross-section and the use of natural target material. Oxygen-14 will be produced as a contaminant and, if proton energies of 11.3 MeV or more are used, also nitrogen-13 will be present. Of the deuteron-induced reactions, the reaction on boron-10 has the advantage of a positive Q-value, however, enriched target material is required. The abundance of ^{10}B is only 20%. This reaction should only be favored above the proton-induced reaction on boron-11 because of limitations enforced by the available accelerator. There are no arguments to use the other deuteron-induced reactions because of the much lower cross-sections and the higher deuteron energies required.

The helium-3-induced reaction on berylium-9 has a positive Q-value and a high cross-section. As with all helium-3 or helium-4-induced reactions the high stopping power for these particles, in comparison with protons or deuterons, will affect the thick target yield substantially. The ^{10}B(^{3}He, pn)^{11}C reaction has to be preferred above the ^{11}B(^{3}He, p2n)^{11}C reaction because the ^{11}C produced via the latter reaction will have a lower yield and will be contaminated with nitrogen-13. The helium-3-induced reaction on carbon-12 allows only for a low specific activity. Also nitrogen-12 and oxygen-14 will be produced as a contaminant. Due to the half-lives of these nuclides they are not a severe problem for most applications. The ^{16}O(^{3}He, 2α)^{11}C reaction has the disadvantage of producing a number of contaminants like ^{18}F, ^{15}O, ^{14}O, and ^{13}N in about the same quantity or more as the desired ^{11}C. All helium-4-induced reactions, as shown in Table 2, have high threshold energies and relatively low cross-sections. Due to this high energy required, the stopping power for the

Table 2
RELEVANT REACTIONS FOR THE PRODUCTION OF ^{13}N

Particle	Reaction	Q-value (MeV)	E(thres.) (MeV)	σ(th.)[a] (mb)	σ(exp) (mb)	Ref.
p	^{13}C(p,n)^{13}N	−3.0	3.2	205	210	114
	^{14}N(p,pn)^{13}N	−10.6	11.3	160	44	119,120
	^{16}O(p,α)^{13}N	−5.2	5.5	120	19	121
					40	122
d	^{12}C(d,n)^{13}N	−0.3	0.35	280	150	122
					110	123
	^{14}N(d,p2n)^{13}N	−12.8	14.6	100		
^{3}He	^{11}B(^{3}He,n)^{13}N	10.2	0	190	5	124
^{4}He	^{10}B(^{4}He,n)^{13}N	1.1	0	340		
	^{11}B(^{4}He,2n)^{13}N	−10.4	14.2	100		

[a] Cross-section calculated according to the statistical model.

helium-4-induced reactions is approximately 1.5 times smaller as for the helium-3 reactions in the energy range of interest. Since the differences in cross-section are approximately a factor of four, the ^{3}He-induced reactions are favored over the ^{4}He-induced reactions. The helium-4 reaction on berylium-9 has the lowest threshold energy and has to be preferred above the reaction on boron-10 because the latter reaction requires enriched target material and also nitrogen-13 will be produced. The helium-4-induced reaction on boron-11 not only requires the highest energy but also shows the lowest cross-section. With this reaction also nitrogen-13 will be produced. The ^{12}C(α,αn)^{11}C reaction allows only for a low specific activity. With this reaction also ^{15}O, ^{14}O, and ^{13}N will be produced.

In conclusion the most promising reactions are the proton-induced reactions on boron-11 and nitrogen-14 and the deuteron-induced reaction on boron-10. The ^{14}N(p,α)^{11}C reaction will have the highest yield followed by the ^{11}B(p,n)^{11}C reaction. The positive Q-value of the ^{10}B(d,n)^{11}C reaction will allow for the production of ^{11}C with rather low energy accelerators.

C. Production of ^{11}C-Labeled Oxides
1. ^{11}C-Labeled Carbon Dioxide

Of the nuclear reactions for the production of carbon-11 the ^{14}N(p,α)^{11}C reaction is used most commonly for ^{11}C-labeled carbon dioxide. The target nuclide is bombarded as nitrogen gas containing trace amounts of oxygen.[61-65] The production can be carried out batch-wise or under continuous flow conditions. For an optimal production yield, the energy absorption in the target gas should be between 18 and 3 MeV, the threshold of the nuclear reaction. The irradiation is normally carried out in a water-cooled aluminum target system containing the gas under pressures of up to 11 bar.[62] In Figure 2 an exploded view of such a target system is shown. Carbon-11 labeled CO_2 is not the primary product in the target gas.[66] Initially ^{11}CN radicals and ^{11}CO are formed. The ^{11}CN radicals are the result of a reaction of recoiling ^{11}C atoms and N_2 molecules, while the ^{11}CO is formed as result of a reaction between ^{11}C and the trace amounts of oxygen present in the target. The radicals and most of the ^{11}CO are oxidized to $^{11}CO_2$ at higher radiation doses. After irradiation the carbon-11 radioactivity is more than 99% $^{11}CO_2$. The only ^{11}C-contaminant is ^{11}CO. The amount of $^{13}N_2$ depends upon the entrance energy on the target gas. From this mixture the $^{11}CO_2$ can easily be separated by a liquid nitrogen cooled trap. When the $^{11}CO_2$ has to be used on line the incident proton energy must be lower than 11.3 MeV the threshold of the ^{14}N(p,pn)^{13}N reaction. The ^{11}CO present can be oxidized to $^{11}CO_2$ by passing the target gas through a furnace containing CuO heated to 800°C. Sometimes extra O_2, up to 4% is added to the N_2 gas. This is not necessary because amounts of O_2 in the order of a few ppm, present in the

FIGURE 2. Exploded view of a high pressure target system for the production of carbon-11.

N_2 gas are sufficient for the oxidation of the ^{11}C-recoil products to $^{11}CO_2$. Addition of extra O_2 even has disadvantages: ^{13}N-labeled nitrogen oxides can be formed by the $^{16}O(p,\alpha)^{13}N$ reaction and nonradioactive nitrogen oxides by radiolysis. When such production conditions are selected, all the ^{13}N-radioactivity is presented as N_2, and the $^{11}CO_2$ can easily be collected in a cold trap. With an incident proton energy of 18 MeV the $^{11}CO_2$ yield is 2mCi/μA.min.[62] At 11.3 MeV, when no $^{13}N_2$ is formed, the yield is 0.7 mCi/μA.min.[67] After prolonged use of the target system the yield is decreasing, presumably because of oxide formation on the inner surface of the system, which traps the $^{11}CO_2$. Treatment of the system with diluted acid followed by H_2O rinsing is then necessary.

In many applications very high specific activities or even carrier-free $^{11}CO_2$ is desired. Theoretically the $^{11}CO_2$ produced is carrier-free. However, in practice, although no carrier is added, the specific activity is far away from carrier-freeness. Several authors have paid attention to this problem.[62,68,69] It is evident that the target system must be designed so that an optimal energy absorption in the target gas is achieved and that the maximum beam current available will result in the highest specific activities. The first source of carrier contamination is caused by impurities in the target gas. The target gas may contain traces of CO_2, but also organic impurities which are oxidized under irradiation conditions to CO_2. So very pure gas, which only contains a trace of O_2, is necessary. In order to reduce the amount of impurities, batch-wise irradiation has to be preferred over continuous flow conditions. Also the volume of the target system must be kept minimal. Another source of carrier contamination is caused by outgassing of the aluminum of the target system and the organic materials used in components like O-rings. Organic materials can be avoided in the design of the system. The effect of outgassing is decreasing after some hours of production under continuous flow. When the target system is stored, it is advisable to keep it pressurized to prevent the leaking of air and consequently of CO_2 and other impurities into the system.

The $^{11}CO_2$ can also be produced by the irradiation of a boron target. However, this method is more complicated than the proton irradiation of molecular nitrogen. The $^{10}B(p,\gamma)^{11}C$ is

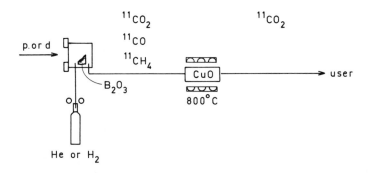

FIGURE 3. Flow scheme for the production of $^{11}CO_2$ by the irradiation of boron oxide with protons or deuterons.

of historical value. Carbon-11 was produced for the first time through this reaction.[70] The chemical form of the target is most often boron trioxyde, either natural[64,71-74] or enriched in ^{10}B.[75] As discussed before, the low threshold energy of the reactions using boron as target nuclide, opens the possibility of using Van de Graaff generators.[76-78] Basically the method works as follows. A glassy B_2O_3 target, obtained by melting powdered B_2O_3 is irradiated under such conditions that the boron oxide just melts and ^{11}C-activity produced can diffuse out. The main ^{11}C-products are ^{11}CO, $^{11}CO_2$, and $^{11}CH_4$. When the beam power is too low to let the B_2O_3 melt, heating by an external source can be applied. For a continuous production the carbon-11 activity is removed from the target system with a sweep gas. After on line purification, the $^{11}CO_2$ is collected in a cold trap. Detailed information on the method and especially on the construction of the target system is reported by the Hammersmith group.[64,72,74] They also investigated the parameters which have influence on the yield such as beam current and energy, sweep gas flow rate, beam distribution, and target material. The product initially formed in the target between ^{11}C-hot atoms and oxygen is ^{11}CO, which is oxidized to $^{11}CO_2$, when no radical scavenger is present. So the composition of the sweep gas influences the product spectrum which is extracted from the target system. When 7.4 MeV protons and helium as sweep gas are used 98% of the activity is found to be $^{11}CO_2$ with a yield of 0.165 mCi/μA.min. The contaminations found are 1.1% ^{11}CO and 0.9% $^{13}N_2$. With 15.6 MeV deuterons and hydrogen as sweep gas the main product is ^{11}CO (86%) together with 8.7% $^{11}CO_2$, 5.2% $^{11}CH_4$ and less than 1% $^{13}N_2$. In this case the ^{11}CO yield is 0.32 mCi/μA.min. To obtain pure $^{11}CO_2$ the ^{11}CO and $^{11}CH_4$ present has to converted into $^{11}CO_2$. This can be achieved on line, when the sweep gas is passed through a CuO furnace at 800°C. After the furnace the $^{11}CO_2$ can be collected from the gas flow by a liquid nitrogen cooled trap. A flow scheme is given in Figure 3. Two other methods are reported in the literature. The first is based on the $^{12}C(\gamma,n)^{11}C$ reaction and can be used when a electron linear accelerator is available.[79] In the second method $^{11}CO_2$ is produced by bombarding CO_2 with 35 MeV protons.[80] The nuclear reaction involved in this method is the $^{12}C(p,pn)^{11}C$ reaction. It is evident that both methods yield a product with a low specific activity.

2. ^{11}C-Labeled Carbon Monoxide

The most convenient way to prepare ^{11}CO is by reduction of $^{11}CO_2$.[64,73] The reduction is carried out on line by passing a $^{11}CO_2$ gas flow through a furnace containing zinc powder heated at 400°C. The reaction equation is

$$Zn + {}^{11}CO_2 \xrightarrow{400°C} ZnO + {}^{11}CO$$

To remove any unconverted $^{11}CO_2$ a tube containing NaOH or LiOH is installed after the zinc furnace.

A direct method using a solid target and the $^{10}B(d,n)^{11}C$ and $^{11}B(d,2n)^{11}C$ reactions is reported by Clark.[64,74] The method is similar to the procedure described for the $^{11}CO_2$ production by bombarding B_2O_3 with deuterons. The only difference is the sweep gas. When hydrogen is used to extract the ^{11}C-products from the target system, 94% of the activity is ^{11}CO, contaminated with about 6% ^{11}C-methane and traces ^{13}N-labeled molecular nitrogen. On line medical application of the ^{11}CO may be limited by these impurities, however, not for instance in the labeling of red blood cells.

D. ^{11}C-Labeled Cyanide

Carbon-11 labeled cyanides are important precursors for the syntheses of ^{11}C-labeled organic compounds. For application the availability of reliable production methods resulting in high yields of ^{11}C-product, is essential. From the literature three different approaches can be extracted.

- An in-target recoil method using the proton bombardment of mixtures of nitrogen and hydrogen.[62,81-83]
- A direct recoil preparation by bombarding a metallic amide or cyanides.[81,82,84]
- Synthetic preparation from a directly produced ^{11}C-labeled precursor.[62,82,85-87]

When mixtures of nitrogen and hydrogen are bombarded with protons at low doses, the main product formed in the target is $H^{11}CN$. The gas mixture may not contain oxygen. Amounts of more than 0.5 ppm O_2 will result in the formation of $^{11}CO_2$ instead of $H^{11}CN$ as the main product.[66] The ^{11}C-labeled cyanide is the result of a recoil reaction between a hot ^{11}C atom obtained by the $^{14}N(p,\alpha)^{11}C$ reaction and nitrogen present in the target. Finn[82] found that the optimum yield was obtained when a mixture of 96.4% N_2 and 5.4% H_2 was bombarded under continuous flow conditions. Radiogaschromatography showed product mixtures containing $H^{11}CN$, $^{11}CH_4$, $^{11}C_2H_2$, ^{11}CO, $^{11}CO_2$ depending on the irradiation conditions. At low radiation doses $H^{11}CN$ is the main carbon-11 product while at higher doses more $^{11}CH_4$ is formed in the target gas. Lamb[81] developed a continuous flow method based on the direct $H^{11}CN$ recovery from the bombarded target gas. The target system used is made of stainless steel, into which a quartz liner is inserted. The quartz liner is essential. From aluminum, brass, or gold-plated target systems it is impossible to recover the cyanide because of absorption onto the walls. Even the quartz liner combined with external heating of the target system does not solve the problem satisfactorily. Only 20% of the theoretically obtainable yield can be extracted as $H^{11}CN$. Christman[62,83] developed a method based on the production of $^{11}CH_4$ directly in the target, followed by on line conversion to $H^{11}CN$.

$$N_2 + H_2 \xrightarrow{^{14}N(p,\alpha)^{11}C} {}^{11}CH_4 \xrightarrow[NH_3]{Pt\ 1000°C} H^{11}CN$$

By following this approach they solved the recovery problem of the $H^{11}CN$. As target gas again an oxygen-free mixture of 94.6% N_2 and 5.4% H_2 is used. The bombardment is carried out under continuous flow conditions at a pressure of 11 bar using an incident proton energy of 18 MeV. The nuclear reaction involved is the $^{14}N(p,\alpha)^{11}C$ reaction. As already mentioned before, the primary product is $H^{11}CN$ together with some ^{11}CO and $^{11}CO_2$ as result of the reaction between recoiling ^{11}C atoms and oxygen impurities present in the target gas. However, at doses greater than 0.1 eV/molecule the $H^{11}CN$ as well as the carbon oxides are reduced to $^{11}CH_4$. The methane is easily recovered from the target system through a copper tubing. A flow scheme is given in Figure 4. To remove any $^{11}CO_2$ present the gas is first passed through a LiOH or NaOH absorber. Next the gas is dried in a dry ice cooled trap and passed into a quartz tube which contains platinum wire heated at 1000°C. In this quartz tube the $^{11}CH_4$ is converted to $H^{11}CN$. The ammonia necessary for this conversion is formed by radiolysis of the target gas. When the beam parameters are not sufficient to form enough

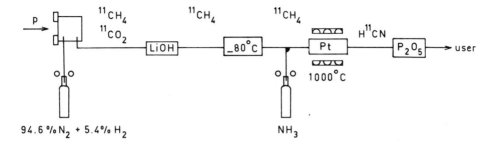

FIGURE 4. Flow scheme for the production of H^{11}CN by the ^{14}N(p,α)^{11}C reaction.

ammonia for a good cyanide yield, NH$_3$ can be added just before the furnace containing the platinum. The unconverted NH$_3$ is separated from the H^{11}CN by P$_2$O$_5$. However, trace amounts may remain present in the final product, which can cause problems when the H^{11}CN is applied. To remove the ammonia, the H^{11}CN can be absorbed in a base resulting in the formation of K^{11}CN or Na^{11}CN. The H^{11}CN can also be collected in a liquid nitrogen cooled trap. The ^{11}CN$^-$ production by this method is rather efficient in comparison with the first mentioned method. Of all the ^{11}C-activity produced in the target 95% can be extracted as ^{11}CH$_4$. The conversion of ^{11}CH$_4$ to H^{11}CN proceeds for 100%. The H^{11}CN yield is about 2 mCi/μA.min. Although no carrier is added in any stage of the procedure the product is not carrier-free. Christman[62] concluded that under this production conditions the initially formed carrier-free H^{11}CN is diluted with stable HCN by a factor of 10^3 to 10^4. Carbon-11 labeled hydrogen cyanide is also formed through the ^{14}N(p,α)^{11}C reaction when lithium amide is irradiated with protons.[81] After bombardment the H^{11}CN is released by dissolving the target in water which contains carrier KCN and a carbonate to scavenge the ^{11}C-oxides which are simultaneously formed in the target. The NH$_3$ formed by hydrolysis of the LiNH$_2$ is removed by boiling the basic solution, and the ^{11}C-labeled carbonate by precipitation. From the solution left the H^{11}CN is recovered by boiling after the addition of H$_2$SO$_4$. The specific activity of the product is low because, for the recovery of the ^{11}CN$^-$, carrier had to be added. The absolute yield achieved is only 3.6% of the theoretical ^{11}C-production.

Finn[82] prepared ^{11}C-labeled metal cyanides by irradiation of the corresponding nonradioactive compound. It is evident that the product obtained by this method is not carrier-free. The ^{11}C-activity is induced in a thick target with 33 MeV protons. With this incident proton energy two nuclear reactions are involved. Firstly the ^{14}N(p,α)^{11}C reaction

$$\text{NaCN} \xrightarrow[\text{}^{12}\text{C(p,pn)}^{11}\text{C}]{\text{}^{14}\text{N(p,α)}^{11}\text{C}} \text{Na}^{11}\text{CN}$$

with a threshold energy of 3.1 MeV and secondly the ^{12}C(p,pn)^{11}C reaction with a threshold of 20.3 MeV. After irradiation the cyanide target pellet can be used directly for synthetic application.

All the synthetic methods for the preparation of H^{11}CN reported start with ^{11}CO$_2$ as precursor. The earliest method is published by Cramer and Kistiakowsky[85] and investigated in detail by Loftfield[86] and Finn.[82] The equation for the reaction is

$$4\text{ K} + \text{NH}_3 + {}^{11}\text{CO}_2 \xrightarrow{620°\text{C}} \text{K}^{11}\text{CN} + \text{KH} + 2\text{ KOH}$$

The conversion is carried out in a sealed quartz ampule at 260°C during 15 min. After the reaction the H^{11}CN is distilled out of the reaction mixture using a vacuum system and H$_2$SO$_4$. For a useful yield it is found necessary to add at least 0.1 mmol carrier CO$_2$. Under these conditions within 30 min after the end of bombardment 90 to 95% of the $^{11}CO_2$ is recovered as $H^{11}CN$. In another synthetic method the $^{11}CO_2$ is reduced to $^{11}CH_4$ by a catalytic method and converted into $H^{11}CN$.[62,87] The reaction sequence is

$$^{11}CO_2 \xrightarrow[370°C]{Ni/H_2} {}^{11}CH_4 \xrightarrow[1000°C]{Pt/NH_3} H^{11}CN$$

The target gas containing the $^{11}CO_2$ and ^{11}CO is mixed with hydrogen and passed through a heated quartz tube containing the nickel catalyst. To remove any unconverted $^{11}CO_2$ the gas stream is then passed over NaOH and dried by $CaSO_4$. The last step is similar to the method of Christman[62] who converts the $^{11}CH_4$ to $H^{11}CN$ by passing the gas stream through a quartz tube containing platinum heated at 1000°C. The overall yield for the conversion of $^{11}CO_2$ to $H^{11}CN$ is 90%.

E. ^{11}C-Labeled Acetylene

Carbon-11 labeled acetylene was first prepared by Cramer and Kistiakowsky[85] by heating ^{11}C-labeled barium carbonate and magnesium. About 60% of the $Ba^{11}CO_3$ can be converted into acetylene. However for a good yield carrier has to be added.[85,88] Myers prepared $Ca^{11}C_2$ directly by bombarding CaC_2 with Helium-3 particles using the $^{12}C(^3He,^4He)^{11}C$ reaction.[89] ^{11}C-acetylene was obtained by hydrolysis of the bombarded carbide, similar to the method used by Cramer. The specific activity of the end-product was rather low; only a few mCi/mmole. A higher specific activity can be obtained through the $^{12}C(p,pn)^{11}C$ reaction.[90] In our laboratory CaC_2 targets are bombarded with 30 MeV protons. This results in a considerable higher yield and specific activity than reported by Myers. However, this product is also far from carrier-free.

From hot atom investigations in the earlier 1960s emerged[11,12] that when cyclopropane is bombarded with protons, ^{11}C-acetylene is formed through the same nuclear reaction. Because of the chemical difference between target and product, very high specific activities can be obtained.[91] Alternatively methane, ethane, or propane can be used as target. Because of the low radiation stability of the alkanes resulting in polymerization, only low beam currents can be used. For routine production of useful amounts of ^{11}C-acetylene this method is not very suitable.[92] Recently Crouzel[93] prepared ^{11}C-acetylene by pyrolysis of $^{11}CH_4$ in an inductive argon plasma. The $^{11}CH_4$ is produced by bombarding a mixture of N_2 and 5% H_2 with protons. The conditions are the same as described for the production of $H^{11}CN$. The $^{11}CH_4$ produced is trapped and introduced batch-wise together with some carrier methane into the base of an argon plasma. The plasma is generated with a high frequency generator using a flow of argon as the plasmagenic gas. Under these circumstances 60% of the $^{11}CH_4$ can be converted momentarily into ^{11}C-acetylene while 40% remains unconverted. After separation, $H^{11}C\equiv CH$ with a purity of 98% and a specific activity of 150 Ci/mmol can be obtained. Of course the specific activity which can be achieved depends upon the available proton beam current.

F. ^{11}C-Labeled Methyl Iodide

Carbon-11 labeled methyl iodide is prepared by two different methods: a synthetic method starting with carbon-11 labeled carbon dioxide[94-96,101] and a direct method by bombarding a mixture of nitrogen and hydroiodic acid with protons.[97,98] The synthetic approach starts with the production of $^{11}CO_2$ and the reduction to $^{11}CH_3OH$. For the last step, the iodination of the methanol,

$$^{11}CO_2 \xrightarrow{LiAlH_4} {}^{11}CH_3OH \xrightarrow{HI} {}^{11}CH_3I$$

the ^{11}C-methanol is released from the complex with lithium aluminum hydride by hydrolysis and carried by a flow of nitrogen gas into hydroiodic acid which is boiled under reflux. Under the conditions used the ^{11}CH$_3$I formed is distilled out of the solution, separated in a soda lime column from evaporating HI and finally dried by passing the nitrogen flow with the methyl iodide through a P$_2$O$_5$ column. The ^{11}CH$_3$I can be collected together with a small amount of ^{11}CH$_3$OH ($< 3\%$) either in a cold trap or by absorbing in an organic solvent. The whole procedure can be performed under remote control and is now in use in many institutes.[100] The conversion of methanol into the iodide is rather fast and efficient. Within 5 to 10 min 80% of the ^{11}CO$_2$ radioactivity can be converted into ^{11}CH$_3$I. The specific activity, however, is much lower than theoretically possible. The two important sources of carrier introduction are the CO$_2$ absorption from the atmosphere on the LiAlH$_4$ and formation of CO$_2$ in the target system by radiolysis of carbon containing impurities. These impurities are released for instance by the metal surface of the system and by carbon containing construction materials like O-rings. When the target system is carefully constructed, outgassed under bombarding conditions and the LiAlH$_4$ used carefully handled, specific activities of several hundreds curie per mmole can be obtained. ^{11}C-labeled methyl iodide can also be prepared by recoil reactions of carbon-11 hot atoms produced by the ^{14}N(p,α)^{11}C reaction in a mixture of N$_2$ and HI. A radiochemical yield of 27% is obtained.[97,98] The other main radioactive gaseous products are ^{11}CO and ^{11}CH$_4$. Advantages of this method are that the ^{11}CH$_3$I is formed directly in the target system and that the specific activity of about 1000 Ci/mmol is much higher than achieved by the synthetic methods starting with ^{11}CO$_2$. However, the absolute yield of ^{11}CH$_3$I is limited. At higher absorbed doses the yield decreases because of radiolytic decomposition. At a fixed beam current the absorbed dose per molecule can be reduced by using a higher flow of target gas through the system. Another problem may be the corrosive nature of the target gas. The production scheme for this method is given in Figure 5. Besides product and radiochemical impurities the bombarded gas at the target outlet contains I$_2$ and H$_2$ formed by radiolysis. The purification of the product is carried out in several steps. At first the HI is removed by bubbling the gas flow through water, while I$_2$ is removed in a Na$_2$S$_2$O$_3$— solution. After drying with sulfuric acid the last traces of I$_2$ are removed in a cold trap. With a proper design no ^{11}CH$_3$I is collected in this trap. From this gas mixture left the ^{11}CH$_3$I can be removed by a cold trap at $-78°$C. For a quantitative absorption a small amount of a suitable solvent like ethanol, ether or THF is introduced into the trap. The system with a flow rate of 100 mℓ/min, an entrance energy of 12.5 MeV and a beam current of 25 μA yields 100 mCi of ^{11}CH$_3$I with a specific activity of more than 1000 Ci/mmol. This amount of radioactivity is obtained during a 1 hr irradiation.

G. ^{11}C-Labeled Methyllithium

Methyllithium is a widely used synthetic intermediate in organic chemistry for introducing methyl groups. Carbon-11 labeled methyllithium is as important as its stable analogue. The preparation of ^{11}CH$_3$Li was first reported by Reiffers[102,103] who prepared the compound by an equilibrium reaction between n-butyl lithium and carbon-11 labeled methyl iodide

$$\text{n-BuLi} + {}^{11}\text{CH}_3\text{I} \rightleftharpoons {}^{11}\text{CH}_3\text{Li} + \text{BuI}$$

The halogen-metal interconversion reaches an equilibrium within a few minutes and according to the law of mass action the excess of n-BuLi which is used, forces the equilibrium towards the ^{11}CH$_3$Li side. To avoid coupling reactions a low temperature is chosen for the interconversion. With an exchange time of 10 min the interconversion is nearly quantitative. The resulting ether solution contains besides ^{11}CH$_3$Li of very high specific activity, a large amount of n-butyl-lithium. Normally this mixture is used for the next synthetic step. It is evident that a mixture of a very high specific active ^{11}C-methyl compound is obtained,

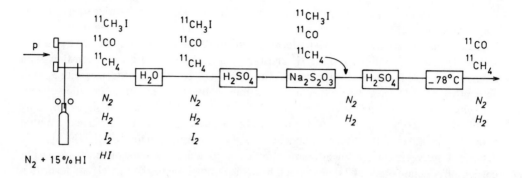

FIGURE 5. Flow scheme for the production of $^{11}CH_3I$ by bombarding a mixture of N_2 and HI with protons. The product is collected in a trap cooled at $-78°C$.

together with a large amount of nonradioactive butylated product. Application in syntheses will of course result in a mixture of the desired ^{11}C-product and the nonradioactive butylated analogue. The labeled compound has to be separated from this mixture. Långström[104] tries to avoid this problem in two ways. Firstly by using t-butyllithium instead of n-butyllithium and secondly by using a polymer supported lithium compound for the interconversion. Rationale for the application of t-butyllithium is the lower reaction rate in following synthetic steps, because of steric hindrance.

H. ^{11}C-Labeled Formaldehyde

Carbon-11 labeled formaldehyde, an important radioactive synthetic precursor, is prepared from $^{11}CO_2$ by reduction to ^{11}C-labeled methanol and subsequent catalytic oxydation to ^{11}C-formaldehyde.[61,69,95,105,106]

$$^{11}CO_2 \xrightarrow{LiAlH_4} {}^{11}CH_3OH \xrightarrow[580°]{catalyst} H^{11}CHO$$

The reduction of the $^{11}CO_2$ to $^{11}CH_3OH$ is carried out by a THF solution of $LiAlH_4$, as described in F. After a complex between $LiAlH_4$ and CO_2 is formed, the THF is evaporated. The $AlLi(OCH_3)_4$ complex is hydrolyzed by water or diluted and the methanol released. Several catalysts are used for the oxidation step. Christman,[61] who introduced the method in the field of carbon-11 chemistry, sweeps the $^{11}CH_3OH$ containing vapor by dry and carbon dioxide free air through a ferri-molybdenum oxide catalyst tube heated at 580°C. Straatmann[105] carried out the oxydation at a lower temperature using iron-molybdenum as catalyst and oxygen as sweep gas. Depending upon the conditions used 50 to 90% of the $^{11}CH_3OH$ can be converted into ^{11}C-formaldehyde. Specific activities up to 170 Ci/mmol are obtainable.[107] The oxidation can also be carried out by silver wool heated at 450°C.[69] Using this catalyst the $^{11}CH_3OH$ is swept with a current of nitrogen gas containing 1 to 20% oxygen, through a quartz tube filled with silver wool. In this way about 60% of the ^{11}C-methanol can be converted. The oxidation process is rather critical. Not only the temperature is important but also the reactivity of the catalyst. The use of a mixture of nitrogen and oxygen is essential because nitrogen gas containing ppm amounts of oxygen deactivates the catalyst with as consequence more unconverted $^{11}CH_3OH$. A lower ^{11}C-formaldehyde yield can also be the result of an overactive catalyst. In that case the desired formaldehyde is further oxidized to carbon dioxide. Because the reactivity of the catalyst is increasing during its use a gradually lower yield of $H^{11}CHO$ is obtained. To overcome this problem preconditioned silver wool can be used which is renewed after every production. An optimum efficiency of the catalyst is obtained when a preoxidation is carried out with stable methanol.[95] Theoretically specific activities of 10^4 Ci/μmole can be obtained. In practice a specific activity of only a few activities of 10^4 Ci/μmole can be obtained. In practice a specific activity of only a few curies per mmol are obtained although no carrier is added. Berger looked systematically to this problem.[69] An important source of carrier contamination is connected with the production of the $^{11}CO_2$. Another way in which undesired carrier is introduced is by the $LiAlH_4$. It is hardly impossible to handle $LiAlH_4$ without any atmospheric CO_2 absorption on the surface. Careful handling and minimizing the amounts of reagents can increase the maximum specific activity. However a too low $LiAlH_4$ concentration in THF may cause that the CO_2 is not completely converted to methanol. Trace amounts of hard to remove THF, left behind on the $LiAlH_4$ surface, are released by the hydrolysis step and cause a lower yield of $H^{11}CHO$. Moreover THF at elevated temperatures in the presence of oxygen and silver wool decomposes into methanol, formaldehyde, and acetaldehyde with a subsequent reduction in specific activity of the end-product. This problem can be reduced by installing a Poropak P trap to hold back the THF.

I. ^{11}C-Labeled Phosgene

Carbon-11 labeled phosgene can be prepared either by a photochemical reaction between ^{11}C-carbon monoxide and chlorine gas[108-111] or by catalytic chlorination of ^{11}CO.[112] The equation for the light induced reaction is

$$^{11}CO + Cl_2 \xrightarrow{h\gamma} {}^{11}COCl_2$$

Principle of the method is the production of $^{11}CO_2$ followed by reduction to ^{11}CO, mixing with Cl_2 and ultraviolet illumination in a quartz ampoula. By selecting the proper conditions radiochemical yields of 100% can be obtained within 5 min after the mixture of ^{11}CO and Cl_2 is transferred into the quartz ampule. The excess Cl_2, necessary for a good yield, can be separated from the ^{11}C-phosgene prepared, by passing the gas after illumination through a tube containing a bed of antimony. The catalytic chlorination of ^{11}CO also starts with the production of $^{11}CO_2$. The produced carbon dioxide collected in a cold trap is swept into a zinc furnace for the conversion into ^{11}CO, using nitrogen as sweep gas. The effluent of the zinc oven is led through a NaOH trap into a tube containing $PtCl_4$ heated at 280°C. Using this procedure radiochemical yields between 60 and 100% are obtained. A flow scheme is given in Figure 6. The $^{11}COCl_2$ is collected from the gas flow by absorbing into toluene. The solubility of unconverted ^{11}CO is too low to be trapped simultaneously. The whole reaction sequence of this rather critical preparation method can be finished within 10 min after the end of the $^{11}CO_2$ collection. Disadvantage of the method is the low specific activity of the end-product. The contamination with carrier probably occurs through the $PtCl_4$. Problems with Cl_2 as consequence of release of chlorine from the heated $PtCl_4$ can be eliminated by purification with antimony.

III. PRODUCTION OF NITROGEN-13

A. Decay of Nitrogen-13

Of the two known positron emitting nuclides of nitrogen, ^{13}N with a half-life of 9.96 m, is the longer-lived.[113] Nitrogen-13 decays by pure positron emission to stable carbon-13 as is shown in the decay scheme in Figure 7. The maximum positron energy is 1.19 MeV and the corresponding range of positrons of this energy in water is 5 mm.

B. Choice of Nuclear Production Reactions

The relevant reactions for the production of nitrogen-13 are shown in Table 2. The proton-induced reaction on carbon-13 has the advantage of a low beam energy.[114] A disadvantage is the need for enriched target material. The use of nitrogen gas as a target material inhibits a high specific activity of the produced nitrogen-13. The required energy for this reaction

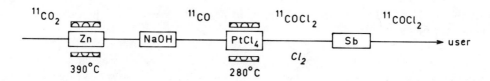

FIGURE 6. Scheme for the conversion of $^{11}CO_2$ into ^{11}C-labeled phosgene.

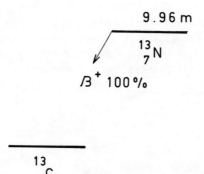

FIGURE 7. Decay scheme of nitrogen-13.

is also well above the threshold of $^{14}N(p,\alpha)^{11}C$ and the $^{14}N(p,n)^{14}O$ reaction. This results in a contamination of the nitrogen-13 by carbon-11 and oxygen-14. The proton-induced reaction on oxygen-16 shows the lowest cross-section, both experimentally and theoretically. In practice, however, clinical useful amounts of nitrogen-13 can be produced via this reaction.[115,116] If natural oxygen is used as target material also some ^{18}F will be present due to the low threshold, 2.5 MeV, of the $^{18}O(p,n)^{18}F$ reaction. The deuteron-induced reaction on carbon-12 allows for a low beam energy with a considerable yield. Also with a Van der Graaff generator useful amounts of nitrogen-13 can be produced via this reaction.[117] There are no arguments, besides machine limitations, to use the deuteron-induced reaction on nitrogen-14. The helium-3-induced reaction on boron-11 has a positive Q-value but the measured cross-section is much smaller than expected from the statistical model. The excitation functions for the helium-4-induced reactions on boron have not been reported. The calculated cross-sections are large but in the calculation of the thick target yields the high stopping power for the helium-4 particles has to be taken into account.

In conclusion the most promising reactions are the proton-induced reactions on carbon-13 or oxygen-16 and the deuteron-induced reaction on carbon-12. The production rates vary from 60 mCi/μA for the $^{16}O(p,\alpha)^{13}N$ reaction to 200 mCi/μA for the $^{13}C(p,n)^{13}N$ reaction.[114,116]

C. ^{13}N-Labeled Nitrates and Nitrites

Nitrogen-13 labeled nitrate and nitrite have been produced for several purposes: for direct application as for instance in denitrification studies or as radioactive precursor for the preparation of ^{13}N-labeled compounds. Of all the nuclear reactions used for the production of ^{13}N as summarized in Table 2, only the $^{16}O(p,\alpha)^{13}N$ reaction yields $^{13}NO_3^-$ and $^{13}NO_2^-$ directly in the target. The most commonly used target materials are H_2O and O_2. When Li_2CO_3 or Al_4C_3 are bombarded with deuterons $^{13}NO_3^-$ and $^{13}NO_2^-$ are also formed directly but with a much lower yield.[117] It is also more difficult to recover the products. But an advantage compared to the (p,α) reaction is the fact that the (d,n) reaction can be used with low energy particle accelerators such as a 3 MV Van de Graaff generator, because of the low threshold of the reaction. Lathrop introduced in 1973 the irradiation of water and oxygen by protons.[125] The irradiations can be carried out either under static conditions or with a continuous flow. For both methods the target system is made of stainless steel or titanium. The system consists of a water cooled plate with a small depression. The depression is covered with a thin stainless steel or titanium foil. The dimensions of the depression depend on the beam parameters. Its volume is mostly between 2 and 10 mℓ. When the production is carried out batch-wise, the target system is filled with distilled water and after irradiation the radioactive solution is pumped into a shielded cell in the radiochemistry laboratory. When very high beam currents are desired the target water can be pumped through the target system to provide additional cooling of foil and target plate.[126] Continuous flow conditions can also adequately be used when for instance low external beam currents make internal irradiation necessary.[127] With an entrance energy of 20 MeV the unprocessed water contains 1.6×10^{-3} mCi ^{18}F and 0.5 mCi ^{15}O per mCi ^{13}N produced.[115] These contaminations are due to the ^{18}O(p,n)^{18}F and ^{16}O(p,pn)^{15}O reactions. When ^{15}O is absolutely not desired in the target, the irradiation has to be carried out below the threshold energy of the ^{16}O(p,pn)^{15}O reaction. To minimize the ^{18}F contamination ^{18}O depleted water can be used as target.[128] The third source of contamination is from recoiled radioactivity produced in the foil which penetrates into the water. Depending on the dose rates in the water ^{13}NO$_3^-$, ^{13}NO$_2^-$, ^{13}NH$_3$, ^{13}N$_2$ and other ^{13}N-labeled gases are formed. Welch and Straatmann[129] found, at dose rates below 1 eV/molecule, 93% ^{13}NO$_2^-$ and 7% ^{13}NH$_3$. At dose rates above 1 eV/molecule 99.7% of the ^{13}N activity was found to be ^{13}NO$_3^-$ and only 0.3% ^{13}NH$_3$. They could not detect any ^{13}NO$_2^-$. Gersberg[130] and Parks[126] confirmed these results. Tilbury and Dahl[131] investigated more parameters involved in the formation of ^{13}N-species. With 14.5 MeV protons they found 5% gaseous radioactivity, 5% ^{13}N-ammonia, 4% nitrite and 86% nitrate. At high dose rates ^{13}NO$_3^-$ is the main product, while at low doses about 40% ^{13}NH$_4^+$ can be obtained. A possible explanation for this effect is that initially ^{13}NH$_3$ is formed in the target by hydrogen abstraction from water. At higher radiation doses a subsequent conversion of the ^{13}NH$_3$ into nitrate and nitrite by radiolysis occurs.[129] For several applications a radiochemical purity of ^{13}NO$_3^-$ produced of better than 99.6% as Parks obtained is adequate and the trace amounts of ^{18}F and other impurities present are not disturbing the measurements. In those cases the bombarded water can be used without any purification. But very often this is not the case, since the maximum dose rates achievable are limited by the accelerator available. Volatile components such as ^{13}N$_2$ and ^{13}N$_2$O can be removed by purging the irradiated water with helium or by evaporation to dryness. When before evaporation the water is made alkaline ^{13}NH$_3$ is also removed and only ^{13}NO$_3^-$ and ^{13}NO$_2^-$ are left behind with ^{18}F and, dependent on the target foil used, trace amounts of other impurities. When nitrogen-13 nitrate is desired the nitrate present can be oxidized by H$_2$O$_2$ after the water is made acidic.[132] Pure labeled nitrite can be obtained by reduction of the nitrate present through a cadmium-copper amalgan column or with a column containing copper dust and acetic acid.[126] High pressure liquid chromatography finally can be used to remove the ^{18}F ions.[127]

D. ^{13}N-Labeled Ammonia

In the literature two methods for the preparation of nitrogen-13 labeled ammonia are reported. The first method is based on the reduction of ^{13}N-labeled nitrates and nitrites formed in a water target by the ^{16}O(p,α)^{13}N reaction as discussed before. In the second method the labeled ammonia is produced directly in a carbon containing target, as consequence of a chemical reaction between the target material and recoiling nitrogen-13 atoms generated by the ^{12}C(d,n)^{13}N reaction. Lathrop[125] introduced the first method. The procedure is as follows: after irradiation the target water is removed remotely from the target system and introduced into a reaction vessel which contains a reducing agent in an alkaline medium. Next the ^{13}NH$_3$ formed is distilled out of the reaction mixture either by steam distillation or by a flow of helium gas. In the first case the steam is condensed and the radioactive ammonia

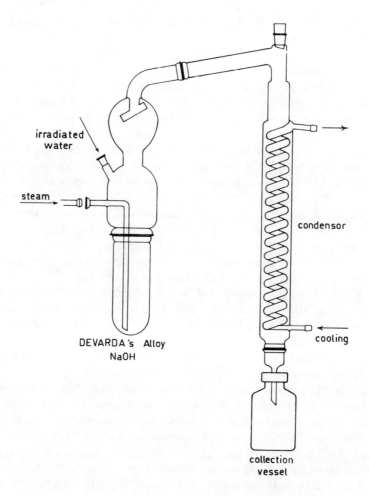

FIGURE 8. Apparatus for the preparation of ^{13}N-labeled ammonia from ^{13}N-nitrate and ^{13}N-nitrite.

solution collected in a few milliliters of water. The distillation process only takes 2 or 3 min. Figure 8 shows a distillation apparatus assembled from glass components normally available in any chemical laboratory. When a flow of helium is used to separate the ^{13}NH$_3$ from the reduction mixture, the reaction flask is heated either by an external source or by the heat generated when the bombarded water is added to the Devarda's alloy and the NaOH pellets. The ^{13}NH$_3$ is collected from the gas flow by passing it through water, saline, or a buffered solution. The whole procedure can be completed within 8 min after the end of bombardment yielding ^{13}NH$_3$ with a radiochemical purity of better than 99.9% when the ^{15}O is not taken into account. Of all the nitrogen-13 activity produced in the target at least 90% can be recovered as ammonia. The reduction of the nitrate and nitrite can be carried out by several reducing agents. Some prefer TiCl$_3$ in alkaline medium while others prefer Devarda's alloy and NaOH[115,132,133] or Ti(OH)$_3$.[134] In the beginning the reduction proceeds rather vigorously and attention has to be paid that no aerosols containing sodium hydroxide enter the condensor. Advisable is to pass the helium flow or the steam through a trap.

Although no carrier is added the product is not carrier-free, mainly because reagents used contain small amounts of ammonium salts or traces of nitrogen compounds which are also reduced to ammonia. To increase the specific activity the reagents can be purified by pretreatment by steam distillation in alkaline medium. Although Krizek[132] takes these precautions she still can detect 7.5 μg carrier in the end-product. An advantage of TiCl$_3$ over

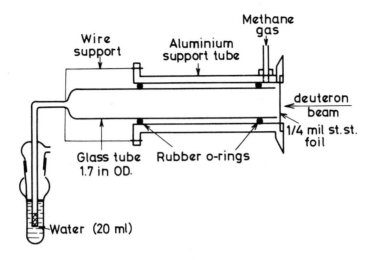

FIGURE 9. Target system for the irradiation of methane gas and bubbler for dissolving $^{13}NH_3$ in water, saline or diluted acid. (From Tilbury, R. S., Dahl, J. R., Monahan, W. G., and Laughlin, J. S., *Radiochem. Radioanal. Lett.*, 8, 317, 1971. With permission.)

Devarda's alloy is that the reagent can be added as solution to the bombarded water. Under remote control the addition of solutions is easier than the addition of solids. Another advantage is that the reduction flask and the distillation equipment can be cleaned without any dismantling and re-used for a next production run.

To obtain useful amounts of $^{13}NH_3$ very often large amounts of radioactivity have to be handled. The only safe way requires remote control or automation. A sophisticated system is described by Ido[135] who obtains the $^{13}NH_3$ by only pushing buttons. All reagents and rinsing solutions are injected through solenoid valves by remotely controlled syringes.

The second $^{13}NH_3$ production method goes through the $^{12}C(d,n)^{13}N$ reaction. Welch[136] showed that ^{13}N-ammonia is formed directly in the target when aluminum carbide is irradiated with 7 MeV deuterons. Also for this nuclear reaction the total radiation dose on the target determines the final chemical product spectrum. At high radiation doses 90% of the ^{13}N-activity was analyzed as $^{13}NH_3$ and about 7% as $CH_3{}^{13}NH_2$. The remaining activity was ^{13}N-labeled cyanide. The irradiated target is processed as follows. The Al_4C_3 is dissolved in diluted HCl, next the solution is made alkaline and the $^{13}NH_3$ released by heating.[137] Tilbury[138] introduced methane as target. The target system, shown in Figure 9 is of the gas flow type and consists of a 100 cm long pyrex glass tube with a diameter of 5 cm. The gas outlet is connected with a bubbler containing water or isotonic saline. The first 10 min of the production is carried out under static conditions using 8 MeV deuterons. The $^{13}NH_3$ is collected while the beam is still on target, by passing the methane through the bubbler. The yield is 4 mCi/μA.15 min and the radiochemical purity is 95%. Less than 3% is $HC^{13}N$, 2% $CH_3{}^{13}NH_2$ and 0.2% $C_2H_5{}^{13}NH_2$. Gelbard[139] investigated the different parameters involved in this production method. Variation in product spectrum was only found by varying the total radiation dose in the target. At high radiation doses $^{13}NH_3$ is the main product. The amount of $HC^{13}N$ is dependent on the flow rate of the target gas. This radiochemical impurity can be reduced to below 0.01% by passing the irradiated gas through a soda lime column. Polymerization in the target gas under irradiation conditions can be troublesome. Even when ultra-high pure methane is used as target gas, a white collodial precipitate appears in the bubbler after some time. The produced $^{13}NH_3$ is not carrier-free although in the procedure no NH_3 carrier is added. Microgram amounts of ammonia could be detected in the solution by Nessler's reagent. Possibly this ammonia is formed by reaction between nitrogen impurities in the target gas and the hydrogen formed by radiolysis.

E. ^{13}N-Labeled Molecular Nitrogen

Procedures using ^{13}N-labeled nitrogen gas belong to the early applications of this radionuclide. In 1940 Ruben[140] described the production of $^{13}N_2$ and demonstrated with this radiotracer the N_2-fixation in plants. Application as radiopharmaceutical was introduced in the early 1960s by the Hammersmith group after Buckingham and Forse made the tracer available as gas[141] and as injectable solution[142] for clinical application. The nuclear reactions which can be used for the production of ^{13}N are compiled in Table 2 together with their nuclear parameters. The ^{12}C(d,n)^{13}N reaction on solid targets such as activated charcoal[140,141,143,145] or graphite[141,144,146] are described by several authors. The radioactivity can be produced under continuous flow conditions or batch-wise by burning the target material after irradiation. Principle of the continuous flow method is the removal of the produced ^{13}N-radioactivity which diffuses out of the solid target, with an appropriate sweep gas. When the target nucleus is bombarded as CO_2, this gas combines both functions of target and sweep gas.[139,144,145,147-151] A very high yield can be obtained by the irradiation of enriched carbon-13 targets.[152] Principle of this batch-wise method is the production of ^{13}N by the ^{13}C(p,n)^{13}N reaction and subsequent combustion of the amorphous carbon target. The third method reported in the literature is based on the ^{16}O(p,α)^{13}N reaction. When an aqueous solution of NH_4Cl is bombarded with protons all the ^{13}N-radioactivity produced, can be collected as $^{13}N_2$ under continuous flow conditions using a sweep gas[153] or by circulating the target solution and collecting the produced $^{13}N_2$.[126] In our laboratory a method for the batch-wise production of $^{13}N_2$ by the same nuclear reaction is developed.[154,155] Principle of this method is the production of ^{13}N-activity with protons on a water target followed by the conversion of the produced ^{13}N-radioactivity into $^{13}NH_3$ and subsequent oxidation to $^{13}N_2$.

1. The $^{12}C(d,n)^{13}N$ Reaction on Solid Targets

The production of $^{13}N_2$ by the bombardment of activated charcoal with deuterons, as well as other methods, are described in detail by Clark and Buckingham[144] in their monograph on short-lived radioactive gases for clinical use. The water-cooled target system they described as well as the air-cooled entrance foil are made of aluminum. In the target system a 0.7 cm layer of activated charcoal granules with a diameter of 2 to 4 mm, is mounted in such a way that under bombardment conditions the ^{13}N-activity produced can diffuse out of the charcoal. Only a part of the ^{13}N-species formed is molecular nitrogen and can, together with trace amounts of impurities, diffuse out of the charcoal. Inherently to this diffusion process the yield of $^{13}N_2$ of this production method is relatively low. The $^{13}N_2$ is recovered from the target system using helium as sweep gas. When this sweep gas is analyzed before any purification, HC^{13}N and C^{15}O are detected as radioactive contaminants. The ^{15}O is produced by the ^{14}N(d,n)^{15}O reaction from the nitrogen which is inevitable as air left behind in the charcoal. In Figure 10 a diagram is given for the purification of the $^{13}N_2$ desired. The HC^{13}N is removed from the sweep gas by a soda lime trap. No provisions are made to eliminate C^{15}O because this impurity is present only in very small amounts. Moreover the amount is decreasing during bombardment because under flow conditions the activated charcoal is gassing out. A typical sweep gas flow rate used by the Hammersmith group[144] is 60 mℓ per minute. With a 35 μA 15.4 MeV deuteron beam on target they measured under continuous flow conditions a production rate of 3.1 mCi per minute. When particles of a higher energy are available higher yields can be obtained.[152] When the $^{13}N_2$ has to be used for pulmonary investigations the ^{13}N-containing sweep gas must be mixed with air for two reasons, firstly to dilute the helium gas to make it tolerable for the patient and secondly to obtain the desired radioactive concentration in the breathing air. Graphite as target material has the advantage over activated charcoal that a much higher yield can be obtained because of a more efficient diffusion of $^{13}N_2$ out of the solid target. Moreover less air and other gases are absorbed in graphite than in activated charcoal. The graphite method can be used

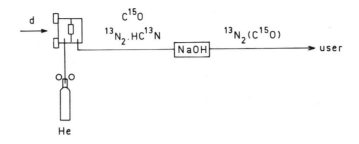

FIGURE 10. Flow scheme for the production of $^{13}N_2$ by the $^{12}C(d,n)^{13}N$ nuclear reaction on activated charcoal.

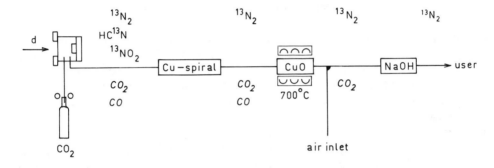

FIGURE 11. Flow scheme for the production of $^{13}N_2$ by the $^{12}C(d,n)^{13}N$ nuclear reaction on graphite. The sweep gas used is CO_2.

for the preparation of $^{13}N_2$ injectable solutions. Again the details of this batch-wise method are described by Clark and Buckingham.[142,144] The material to be bombarded is a block of graphite. To make the diffusion more efficient the block of graphite is carved horizontally and vertically to a depth of 8 mm into prisma's with a cross-section of a few millimeters. The graphite is contained in a water-cooled aluminum target system. Carbon dioxide is used as sweep gas for several reasons. Heated by the beam the graphite must reach a temperature of 1450°C. At this temperature the block reaches white heat resulting in a higher diffusion rate of the $^{13}N_2$. Simultaneously a second process is proceeding in which the CO_2 plays a key role. The block of graphite is eroded slowly by the oxidation of carbon through the C + CO_2 → 2CO reaction, which has positive influence on the recovery of the $^{13}N_2$ from the graphite. Nevertheless, only 6% of the ^{13}N produced in the target can be recovered as $^{13}N_2$. It is obvious that for a useful yield the bombarding conditions are quite critical. High power densities are needed to heat the graphite. After the produced $^{13}N_2$ is removed from the target system it has to be purified. Because of the erosion process the sweep gas contains substantial amounts of nonradioactive CO. This makes it not advisable to use this $^{13}N_2$ on line for physiological investigations. Only two radiochemical impurities are detected in the gas flow: $HC^{13}N$ and $^{13}NO_2$. In Figure 11 a flow diagram including purification traps for continuous production is given. The $HC^{13}N$ and $^{13}NO_2$ are removed by a small bore copper tubing at room temperature in which the compounds are absorbed. The CO is converted into CO_2 in the copper oxide furnace. For production under continuous flow, after the copper oxide furnace the sweep gas is mixed with air and passed through a NaOH column to remove the CO_2. In the batch-wise production mode and without diluting the sweep gas with air specific activities can be obtained which are suitable for the preparation of $^{13}N_2$-solutions. However high beam currents are necessary. Clark and Buckingham obtain, of course under their particular bombarding conditions of 60 μA 15.4 MeV deuterons, a $^{13}N_2$ yield of 0.4 mCi/

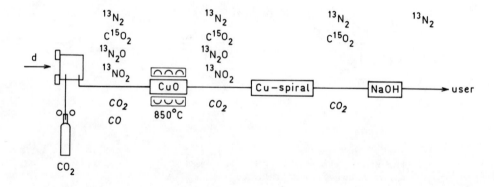

FIGURE 12. Flow scheme for the production of $^{13}N_2$ by the $^{12}C(d,n)^{13}N$ reaction and CO_2 as target gas.

μA min. From the activity produced they prepared a solution of $^{13}N_2$ in saline with a radioactivity of 0.21 mCi/mℓ. Ritchie[146] also used graphite. His target system is less sophisticated than the Hammersmith target. With a 200 μA 3 MeV deuteron beam, accelerated by a Van de Graaff generator, he reports a yield of 0.02 mCi/μA.min. Of course the much lower deuteron energy is the main reason for a lower yield, but a part of the discrepancy can be explained by a less efficient $^{13}N_2$ diffusion out of the graphite. Because of the low maximum energy of Van de Graaff generators, in this case 3 MeV, energy absorption in foils can reduce the yield dramatically. To avoid this problem Ritchie described a foil-less separation between the vacuum of the accelerator and the sweep gas in the target. The principle of this method is a small aperture and differential pumping to prevent leakage of sweep gas into the accelerator.

2. The $^{12}C(d,n)^{13}N$ Reaction on Carbon Dioxide

The production method of $^{13}N_2$ by the bombardment of carbon dioxide gas can also be done batch-wise or under continuous flow conditions. The target system is a water-cooled aluminum or brass holder with an air-cooled beam entrance foil which reduces the deuteron energy to 6.5 MeV. The ^{13}N is induced in the target gas by the $^{12}C(d,n)^{13}N$ reaction. Although carbon dioxide only contains 27% of the carbon target nuclide the very low toxicity of the gas and the easy separation of $^{13}N_2$ and CO_2, makes CO_2 very suitable for production. Welch[147] investigated the fate of ^{13}N hot atoms in different mixtures of CO_2 and N_2. Under bombardment with deuterons at low radiation doses a mixture of $^{13}N_2$, $^{13}N_2O$ and ^{13}NO is formed. At higher radiation doses the nitrous oxide peak decreases because of radiolysis. Clark and Buckingham[144] found as products, when analytical grade CO_2 gas was irradiated, more than 99% $^{13}N_2$ contaminated with trace amounts of $^{13}N_2O$, $^{13}NO_2$ and $C^{15}O_2$ from the $^{14}N(d,n)^{15}O$ competing reaction. In Figure 12 a flow scheme is given for the purification of $^{13}N_2$ produced. After irradiation the bombarded CO_2 is passed through a CuO furnace at 850°C to convert the CO eventually formed from CO_2 by radiolysis back into CO_2, and through a spiral in which $^{13}NO_2$ is removed. The CO_2 is absorbed in a soda lime column. The gas left can be used directly when mixed with air for pulmonary investigations or for the preparation of injectable solutions of $^{13}N_2$ in saline. When the production method is intended to make $^{13}N_2$ solutions CO_2 with a very low amount of N_2 has to be used as target gas.

3. The (p,n) Reaction on Enriched Carbon-13

When the reported cross-sections of the $^{12}C(d,n)^{13}N$, the $^{16}O(p,\alpha)^{13}N$ and the $^{13}C(p,n)^{13}N$ reactions in Table 2 are compared, it is clear that the last nuclear reaction has to be preferred for production. When only low beam currents are available the (p,n) reaction can provide very useful amounts of ^{13}N-radioactivity. The batch-wise preparation of ^{13}N by this nuclear reaction is described in detail.[152,156] The method is not advised for making ^{13}N$_2$ solutions. The ^{13}N-radioactivity is produced by proton bombardment of solid amorphous carbon-13 targets enriched to 97%. The carbon is contained in a gas-tight target system to prevent that gaseous radioactive products which might be formed during irradiation, are released. After irradiation the target is removed from the target system and combusted by an automated Dumas combustion procedure. The combustion is carried out in a quartz tube after mixing the carbon with CuO powder and KNO$_3$. The CuO provides the oxygen for the combustion and KNO$_3$ is added as source for carrier nitrogen gas. The combustion products are swept with CO$_2$ through heated tubes filled with CuO and Cu. In the first tube nitrogen oxides are formed which are reduced to ^{13}N$_2$ in the second tube. The sweep gas is removed from the ^{13}N$_2$ by a liquid nitrogen trap. This trap has a second function: the reduction of the concentration of nitrogen oxides. Because the addition of KNO$_3$, to provide carrier N$_2$ which is necessary for reproducible results, the final volume of the labeled nitrogen gas is 0.02 mℓ which makes the method unsuitable for preparing ^{13}N$_2$ solutions. The ^{13}N$_2$ gas is delivered batch-wise within 15 min after the end of bombardment. With irradiation times of 20 to 30 min up to 20 mCi/µA can be produced when the bombardment is carried out with 11 MeV protons. NO was the only chemical impurity in the final product as was detected by ^{15}N-mass spectrometry. The ^{15}N was added as K^{15}NO$_3$. Decay curve analysis showed that the ^{13}N$_2$ is essentially free of radionuclide impurities.

4. The (p,α) Reaction on Water

The ^{16}O(p,α)^{13}N reaction can be used for batch-wise production of ^{13}N$_2$ gas.[126,127,153,154] as well as for the preparation of ^{13}N$_2$-injectable solutions.[155] Principle of this method is the reduction of ^{13}N-labeled nitrate to ^{13}NH$_3$, followed by oxidation of the ammonia to molecular nitrogen. The main advantages of this method over other methods are the high absolute yield of ^{13}N$_2$ and the very high amount of radioactivity per mℓ of nitrogen gas or per mℓ injectable ^{13}N$_2$ solution. Moreover the convenient transportation of the radioactivity as ^{13}NH$_4^+$ from the radiochemistry laboratory to the clinical facility and the simplicity of generating gas immediately before use, increases the application flexibility. The ^{13}N-ammonia is produced by bombarding water as described before. When the ^{13}NH$_3$ is collected in a vessel containing carrier ammonium salt and after an oxidation agent like sodium hypobromite is added, ^{13}N$_2$ gas is formed immediately and escapes from the solution. Of course the radioactivity per mℓ N$_2$ gas depends upon the amount of carrier added. The ^{13}N$_2$ can be used directly for inhalation or swept for instance into a spirometer system. When the aim is to prepare a sterile injectable solution no carrier is added to the ammonia collected and the fraction of the distillate with the highest amount of radioactivity is used. To achieve a radiochemical purity of more than 98% the oxidation is carried out with an excess of NaOBr. NaOBr is not acceptable in an injectable solution and removal of the excess is necessary. This is done by adding ascorbic acid which converts the hypobromite into bromide as suggested by Krizek.[157]

$$\text{OBr}^- + \text{ascorbic acid} \longrightarrow \text{Br}^- + \text{dehydroascorbic acid}$$

The high amount of ^{13}N$_2$ per mℓ solution obtained by NaOBr oxidation is achievable because the ^{13}N$_2$ is generated in the solution, while in the other methods the ^{13}N$_2$ is generated first and then forced into solution. Suzuki[153] investigated recoil reactions of ^{13}N hot atoms generated by the ^{16}O(p,α)^{13}N reaction in solutions and found that irradiation of aqueous ammonia results in the production of ^{13}N$_2$ with a radiochemical purity of better than 99.9%. This method is not suitable for the preparation of ^{13}N$_2$ solutions. For this application the specific activity of the ^{13}N$_2$ gas is too low because of N$_2$ carrier formation by radiolysis of the NH$_3$ in the target solution. He reports with a 1 µA proton beam for 5 min a specific activity of 3.5 mCi/mℓ of gas.

F. ^{13}N-Labeled Nitrogen Oxides

Starting with ^{13}N-labeled nitrate Nickles[156] developed a method for the preparation of ^{13}N$_2$O, an inert diffusable tracer which can be used to investigate cerebral blood flow. The ^{13}N$_2$O is formed from NH$_4$^{13}NO$_3$ or ^{13}NH$_4$NO$_3$ by heating with sulfuric acid at 260°C.

$$NH_4NO_3 \longrightarrow N_2O + 2H_2O$$

The ^{13}N radioactivity is produced by bombarding water. When ^{13}NH$_4$NO$_3$ is used as intermediate, the ^{13}NO$_3^-$ formed in the water is reduced to ^{13}NH$_3$ followed by the addition of carrier NH$_4$NO$_3$. When the nitrate labeled analogue is used as intermediate carrier NH$_4$NO$_3$ is added directly to the bombarded water. Both methods yield ^{13}N$_2$O which is contaminated with about 30% ^{13}N$_2$. It was not possible to detect any ^{13}NO or ^{13}NO$_2$. When a higher excess of NH$_4^+$, added as (NH$_4$)$_2$SO$_4$ to the pyrolysis mixture, is used the ^{13}N$_2$ radiochemical contamination can be reduced to 2%, however, at cost of a lower specific activity. With a preparation time of 20 min about 80% of the initially available ^{13}NO$_3^-$ can be converted into ^{13}N$_2$O.

Nitrogen dioxide is prepared by Ogg.[158] The ^{13}N radioactivity is produced by the ^{12}C(d,n)^{13}N reaction on graphite. After irradiation the target is burned in a flow of oxygen and the oxides formed are collected at liquid nitrogen temperature in a trap which contains carrier NO$_2$. The ^{13}NO$_2$ was recovered from the trap with carbon tetrachloride.

IV. PRODUCTION OF OXYGEN-15 AND OXYGEN-14

A. Decay of Oxygen-15 and Oxygen-14

Of the three known positron-emitting nuclides of oxygen the longest-lived is oxygen-15 with a half-life of 122 s. This radionuclide decays by positron emission for 99.9% and electron capture for 0.1% to stable nitrogen-15.[159] The decay scheme is shown in Figure 13. The maximum positron energy is 1.72 MeV, corresponding to a range of 8 mm in water.

Oxygen-14 with a half-life of 70.6 s decays to stable nitrogen-14. The decay of oxygen-14 is a pure positron decay[160] as shown in the decay scheme in Figure 14. In most cases, 99.3%, oxygen-14 decays to the 2.3 MeV level in nitrogen-14. In this case the maximum positron energy is 1.81 MeV, corresponding to a range of 8.4 mm in water. In 0.6% of the cases the oxygen-14 nucleus decays directly to the ground state of nitrogen-14. The maximum positron energy is then 4.12 MeV and the corresponding range in water is 21 mm. In a very few cases, 0.06%, oxygen-14 decays to the 3.9 MeV level in nitrogen-14. The decay of oxygen-14 is characterized by annihilation radiation and by a γ-ray of 2.313 MeV.

B. Choice of Nuclear Production Reactions

The relevant cross-sections for the production of oxygen-15 are summarized in Table 3. Experimentally only the excitation functions for the two most promising reactions have been measured. The increase in cross-section in the course of time for the ^{14}N(d,n)^{15}O reaction is remarkable. The proton-induced reaction on ^{16}O has a much lower cross-section, both experimentally and theoretically, than the deuteron-induced reaction on ^{14}N. Since the stopping power for protons on oxygen is approximately a factor 5 lower than the stopping power for deuterons on nitrogen, in the energy ranges of interest, the thick target yield of the ^{16}O(p,pn)^{15}O reaction will not be as low as suggested by the difference in cross-section. The use of oxygen as a target material inhibits a high specific activity. There are no arguments to use the deuteron-induced reaction on oxygen because of the required deuteron energy and the cross-section calculated. The ^{13}C(^{3}He,n)^{15}O reaction has the advantage of a positive Q-value but the need for enriched target material is a disadvantage. The ^{4}He-induced reaction on ^{12}C has a high cross-section according to the statistical model calculation, but the high

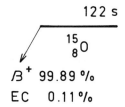

FIGURE 13. Decay scheme of oxygen-15.

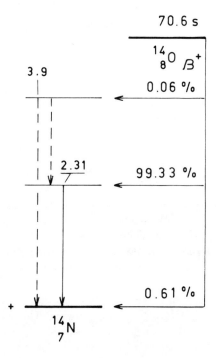

FIGURE 14. Decay scheme of oxygen-14.

stopping power for ^{4}He particles has to be taken into account for the calculation of the thick target yield.

The reactions reported in the literature for the production of ^{15}O are the ^{16}O(p,pn)^{15}O reaction[161] and the ^{14}N(d,n)^{15}O reaction.[162,164,165] If the energy of the deuteron beam is kept below 10 MeV no ^{14}O and ^{13}N will be produced. The energy of the proton beam should be less than 30 MeV in order to avoid ^{14}O, but ^{13}N will always be produced.

The reactions for the production of ^{14}O (T1/2 = 70.6 s) are summarized in Table 4. Since no excitation functions are available for the different reactions the cross-section has been calculated according to the statistical model. For the proton-induced reactions the reaction on ^{14}N has for production all the advantages like a high yield, low beam energy and no contamination with oxygen-15, since the cross-section of the ^{14}N(p,γ)^{15}O reaction is very

Table 3
RELEVANT REACTIONS FOR THE PRODUCTION OF ^{15}O

Particle	Reaction	Q-value (MeV)	E(thres.) (MeV)	σ(th.)[a] (mb)	σ(exp) (mb)	Ref.
p	$^{16}O(p,pn)^{15}O$	−15.7	16.6	165	75	166,167
d	$^{14}N(d,n)^{15}O$	5.1	0.0	310	28	168,169
					230	165
					300	170
	$^{16}O(d,p2n)^{15}O$	−17.9	20.1	110		
^{3}He	$^{13}C(^3He,n)^{15}O$	−8.5	11.3	385		

[a] Cross-section calculated according to the statistical model.

Table 4
RELEVANT REACTIONS FOR THE PRODUCTION OF ^{14}O

Particle	Reaction	Q-value (MeV)	E(thres.) (MeV)	σ(th.)[a] (mb)
p	$^{14}N(p,n)^{14}O$	−5.9	6.3	100
	$^{16}O(p,p2n)^{14}O$	−28.9	30.7	13
	$^{14}N(d,2n)^{14}O$	−8.2	9.4	35
	$^{16}O(p,3n)^{14}O$	−31.1	35.0	6
^{3}He	$^{12}C(^3He,n)^{14}O$	−1.1	1.4	120
^{4}He	$^{12}C(^4He,2n)^{14}O$	−21.7	28.9	33

[a] Cross-section calculated according to the statistical model.

low. A contamination with ^{11}C will always be present since the threshold for the $^{14}N(p,\alpha)^{11}C$ is 3.1 MeV. The maximum in the cross-section is reached at 16 MeV. At this energy also the threshold for the $^{14}N(p,pn)^{13}N$ reaction is reached. ^{13}N will then be produced as a contaminant. There are no arguments to use a deuteron-induced reaction since the deuteron energies required are higher than the required proton energies, while the calculated cross-sections are substantially lower. The ^{3}He-induced reaction on ^{12}C has also a high cross-section. The difference in stopping power, however, between protons and ^{3}He-particles in the energy ranges of interest, is approximately a factor six. This will result in a much lower thick target yield for the ^{3}He-induced reaction. A technical problem for the ^{3}He and ^{4}He-induced reactions, is a fast recovery of the ^{14}O produced. The ^{4}He-induced reaction not only has a rather low cross-section but also requires a much higher beam energy. The only reaction used for the production of ^{14}O is the $^{14}N(p,n)^{14}O$ reaction.[163,164]

C. ^{15}O-Labeled Molecular Oxygen

As discussed before, the most commonly used nuclear reaction for the production of oxygen-15 labeled oxygen gas is the $^{14}N(d,n)^{15}O$ reaction. Apart from nuclear physics arguments the deuteron-induced reaction is the best for two reasons: natural nitrogen gas can be used as target and even Van de Graaff generators are able to produce useful amounts of ^{15}O with this reaction. Moreover the inertness of the N_2 which is used simultaneously as target and sweep gas, is an advantage in medical application, particularly when the radioactivity is administered to the patient on line by breathing.

For production most often a flow of N_2 gas containing 1 to 4% O_2 is bombarded. The deuteron energy absorbed in the target gas should be between 6.5 and 3.0 MeV[171-174] to avoid the production of contaminating radionuclides. The ^{15}O-labeled O_2 is formed directly

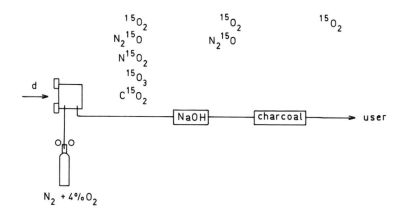

FIGURE 15. Flow scheme for the production of $^{15}O_2$.

in the target by a hot atom reaction between recoiling ^{15}O atoms and the O_2 present. The rate constant of this hot atom reaction must be higher than the rate constant of the reaction between hot oxygen atoms and hydrogen or other impurities present in the target gas.[172] It is obvious that the product formed is not carrier-free. The product spectrum formed directly in the target is investigated in detail.[172,173] Clark and Buckingham found that when N_2 gas with a chemical purity of 99.9% is irradiated with 6.3 MeV deuterons, most of the ^{15}O-radioactivity is recovered from the target system as $N^{15}O_2$ and ^{15}O-labeled ozone. Only 35% was ^{15}O-labeled O_2. Minor contaminants found are $N_2^{15}O$ and ^{15}O-labeled CO_2. However, when a mixture of N_2 and O_2 is bombarded a total different product spectrum is formed. With a mixture of N_2 and 4% O_2, 97% of all radioactivity recovered from the target is $^{15}O_2$ and only 2.2% $N_2^{15}O$. Only trace amounts of other impurities such as $C^{15}O_2$, $N^{15}O_2$, $^{15}O_3$, $C^{15}O$ and $^{13}N_2$ are formed. Carbon monoxide and carbon dioxide are probably the result of the reaction between $^{15}O_2$ and trace amounts of organic impurities present in the target gas and the target system. Also the radiation dose on the target gas influences the product spectrum. When the beam current is increased from 5 to 50 μA the $N_2^{15}O$ impurity decreased from 2.2% to below 1% with a simultaneous increase of the purity of the $^{15}O_2$ to above 99%. For many applications this purity is sufficient. For medical application, however, the risk that the purity decreases because of variation in beam conditions is not acceptable. In Figure 15 the production scheme is given for the continuous production of pure $^{15}O_2$. The target system is fed from a tank containing the nitrogen-oxygen mixture. Depending upon the construction of the target system, the gas can be bombarded under pressure in order to reduce the length of the target system. After the gas has left the target system it is passed through a NaOH absorber in which $C^{15}O_2$ and $N^{15}O_2$ are removed and in which the ^{15}O-labeled ozone, eventually present, is decomposed to $^{15}O_2$. The $N_2^{15}O$ is absorbed in a charcoal column. Because of convenience an open system is preferred over a closed system in which a fixed volume of gas is recirculated through the target during bombardment.[174] To increase the specific activity the amount of O_2 in the target gas can be reduced to 0.5%. When a recirculating system is used minimizing the volume of target and the processing system can also be helpful. When for some reason, only a proton beam with a relatively high energy is available or when switching from protons to deuterons has to be avoided the $^{16}O(p,pn)^{15}O$ reaction on oxygen gas can be used.[175] However, the simultaneous production of ^{13}N by the competing $^{16}O(p,\alpha)^{13}N$ reaction can be troublesome. When only the high energy part of the excitation function is used, the amount of ^{13}N can be kept below acceptable levels. Under the circumstances Beaver and co-workers are using, an irradiation of 5 min with a beam current of 15 μA, an amount of 30 to 35 mCi is collected in a syringe. The amount of radioactivity is measured at 1 min after EOB.

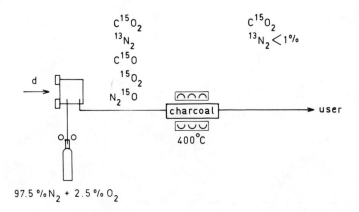

FIGURE 16. Flow scheme for the production of $C^{15}O_2$.

D. ^{15}O-Labeled Carbon Dioxide

Oxygen-15 labeled CO_2 can be produced in two different ways, both using the $^{14}N(d,n)^{15}O$ nuclear reaction. In the first method $^{15}O_2$, produced as already described, is converted to $C^{15}O_2$ with activated charcoal at 400°C.

$$C + {}^{15}O_2 \xrightarrow{400°C} C^{15}O_2$$

The conversion is achieved continuously in a furnace with an activated charcoal column in the gas line coming from the target system. In the second method a mixture of CO_2 and N_2 is bombarded with 6.3 MeV deuterons. Under these production conditions the $C^{15}O_2$ produced is formed directly in the target system probably by reaction of highly energetic ^{15}O-atoms. Some nonradioactive CO is formed by radiolysis of CO_2. Clark and Buckingham investigated both methods extensively.[173] For the in-target production method a mixture of 2.5% CO and 97.5% N_2 is irradiated with 6.3 MeV deuterons under slightly elevated pressure to keep the gas flowing. Analysis of the target gas by radiogaschromatography showed that $C^{15}O_2$ was the main product with only trace amounts of $C^{15}O$, $^{15}O_2$, $N_2^{15}O$ and $^{13}N_2$ as contaminants. By installing an activated charcoal furnace kept at 400°C in the flow system, see Figure 16, the ^{15}O-labeled impurities can be removed. The $^{13}N_2$ activity was less than 1%.

E. ^{15}O-Labeled Carbon Monoxide

The only method found in the literature for the production of $C^{15}O$ is through $^{15}O_2$ by the $^{14}N(d,n)^{15}O$ reaction as described before and the conversion of the $^{15}O_2$ by activated charcoal at 900°C.

$$C + {}^{15}O_2 \xrightarrow{900°C} 2\, C^{15}O$$

Details of the method can also be found in the monograph by Clark and Buckingham.[173] As target and sweep gas a mixture of N_2 and 2% O_2 is used. Under continuous flow conditions this mixture is bombarded with 6.3 MeV deuterons. The product spectrum of the bombarded gas at the output of the target system is 96% $^{15}O_2$, 2.9% $N_2^{15}O$ and trace amounts of $C^{15}O$, $C^{15}O_2$, $N^{15}O_2$, $^{15}O_3$ and $^{13}N_2$. The flow scheme is given in Figure 17. After this mixture has been passed through an activated charcoal furnace at 900°C the radiochemical purity of the product is better than 99%. A soda lime trap can be installed to absorb any $C^{15}O_2$ from the gas flow.

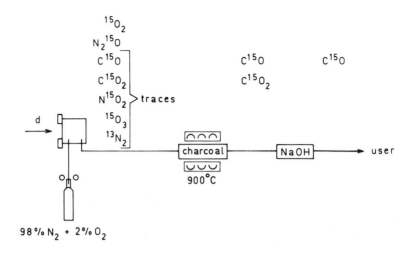

FIGURE 17. Flow scheme for the production of $C^{15}O$.

F. ^{15}O-Labeled Water

Three different methods are found in the literature for the production of $H_2^{15}O$. In the first method $^{15}O_2$ is produced as described before, mixed with H_2 and converted into $H_2^{15}O$ by a catalyst.[164,173] Deoxo-palladium-coated alumina pellets heated to 150°C are used, and as target gas a mixture of O_2 and N_2. Eventually formed $C^{15}O_2$ can be removed by a sodium hydroxide trap. The second method makes use of $H_2^{15}O$ formation directly in the target.[176,177] In this case a mixture of 5% H_2 in N_2 at a flow of 100 mℓ/min is bombarded with 3.5 MeV deuterons. The ^{15}O is produced via the $^{14}N(d,n)^{15}O$ reaction and water vapor is formed in the target system by a hot atom reaction of recoiling ^{15}O atoms and H_2. The $H_2^{15}O$ is not carrier-free because it is impossible to remove all carrier H_2O from the target system. The labeled water formed is swept out of the target system and is trapped continuously on a glass bed. By pumping sterile saline over the glass bed an injectable $H_2^{15}O$ saline solution is obtained. No significant radiochemical or radionuclidic impurity was found in the final product. The third method[178] makes use of the fast exchange between ^{15}O-labeled carbon dioxide and carbonic acid.

$$H_2O + CO_2 \rightleftharpoons H_2CO_3$$

The exchange reaction is carried out by passing $C^{15}O_2$ into water. In blood the exchange reaction is much faster than in water, due to the enzyme carbonic anhydrase. Within 30 sec the exchange is virtually quantitative; 99.99% of the radioactivity appeared as water.

G. ^{14}O-Labeled Molecular Oxygen and ^{14}O-Labeled Water

The production of ^{14}O as oxygen gas is described by Dahl and co-workers.[163] Nickles[164] reported on the catalytic conversion of the ^{14}O-labeled O_2 into $H_2^{14}O$. The systems used are designed for production of steady state amounts of radioactivity.[179] Because production through a gas flow system is more convenient than bombarding a solid target material the $^{14}N(p,pn)^{14}O$ reaction is preferred over nuclear reactions on carbon-12. Competing nuclear reactions are $^{14}N(p,pn)^{13}N$ with a Q-value of $-$ 8.3 MeV and the $^{14}N(p,\alpha)^{11}C$ reactions with a Q-value of $-$ 2.8 MeV. Depending upon the entrance energy of the protons on the target gas and the energy degradation in the target ^{13}N and ^{11}C are produced as contaminants in substantial amounts. Nitrogen-13 can be avoided by reduction of the proton entrance energy below the threshold of the $^{14}N(p,pn)^{13}N$ reaction. It is impossible to avoid the simultaneous production of carbon-11. By irradiation with 8.5 MeV protons for instance,

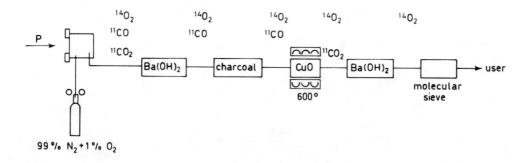

FIGURE 18. Flow scheme for the production of $^{14}O_2$.

approximately equal amounts of ^{14}O and ^{11}C are formed. Pure ^{14}O by this method can only be obtained after on line purification methods. Dahl[163] described in detail the production of $^{14}O_2$ using 8.5 MeV protons on nitrogen containing 1% O_2. The energy absorption in the target is 1.5 MeV. As consequence of the added oxygen to the target gas the product is not carrier-free. In Figure 18 a scheme is given of the continuous flow system. Because of the short half-life of ^{14}O a careful design of the target and the flow system is necessary to minimize losses of radioactivity by decay. A pressurized target system in combination with small diameter tubing has to be used to obtain an optimum yield. After irradiation the target gas is purified by a sequence of absorbers. $Ba(OH)_2$ is used for removal of $^{11}CO_2$ and nonradioactive NO_2, activated charcoal for organic impurities and ozone. Next the gas flow is passed through a tube containing CuO heated at 600°C. In this tube ^{11}CO is oxidized to $^{11}CO_2$. This $^{11}CO_2$ is removed by another bariumhydroxide column. With an energy degradation of 1.5 MeV and a flow rate of 1.3 ℓ/min a yield of 0.12 mCi/μA.min can be obtained. The purity of the $^{14}O_2$ is 99.9% with ^{11}C as only radionuclidic impurity left.

Oxygen-14 labeled water can be prepared on line by the catalytic conversion of $^{14}O_2$. The only method described in the literature[164] starts with the production of $^{14}O_2$ under continuous flow conditions by bombarding a thick nitrogen target containing 400 ppm O_2 as carrier. After bombardment the target is mixed with H_2 and O_2 and the $^{14}O_2$ converted quantitatively to $H_2^{14}O$ by passing the gas mixture over a platinum-on-alumina catalyst contained in a water-cooled quartz tube. Because the exothermal reaction is difficult to control it is preferable to use a platinum-on-alumina catalyst heated to about 150°C as described for the production of $H_2^{15}O$.[173]

V. LABELING ORGANIC COMPOUNDS WITH CARBON-11 AND NITROGEN-13

A. The Literature

Because potentially any organic or biochemical reaction can be used for labeling compounds with carbon-11 and nitrogen-13, it is impossible to give an overview of all the syntheses useful as preparation method. When searching the literature the chemist faces the problem that only in a very few cases the optimum reaction time is included in which the reported yield is obtained: every reaction has to be assessed or tried out on a useful radiochemical yield within the time limit set by the short half-lives of the nuclides.

In the preceding part of this chapter the methods to produce short-lived labeled synthetic precursors are discussed. Here the preparation of carbon-11 and nitrogen-13 organic compounds, starting with any of these radioactive precursors will be discussed. For carbon-11 the literature up to 1951 is adequately reviewed by Calvin[108] and by Kamen.[180] Wolf and Redvanly[8] cover the literature until May 1976. For nitrogen-13 labeled radiopharmaceuticals a review until May 1976 by Straatmann[9] is available. For this review we looked after synthetic

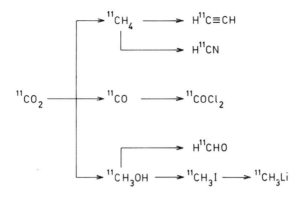

FIGURE 19. Synthesis of simple ^{11}C-compounds from $^{11}CO_2$.

labeling methods published between 1975 and 1981. Although not all reports in this period contain essential new synthetic information, they are still included. Many articles do not contain relevant information on yield, specific activity, or preparation time, which omission makes them less useful as they could be.

It is desirable not only to report on the amounts of chemicals used and the different steps between nuclide production and purified end-product, but also on the time necessary to reach the different stages of the procedure. As reference time it is convenient to use EOB. Detailed information on pretreatment of starting materials and other chemicals or biochemicals used, and on the pretreatment of the glassware can also be very helpful in repeating the procedure.

B. Carbon-11 Labeled Organic Compounds

The most commonly used ^{11}C-precursors are $^{11}CO_2$ and H^{11}CN. In Figure 19 a scheme is given of simple ^{11}C-compounds which can be prepared from $^{11}CO_2$, either on line or batch-wise, as discussed before.

Other precursors used are ^{11}CO, H^{11}C≡CH and recently ^{11}C-guanidine produced by proton irradiation of a liquid NH_3-N_2O[181] or ammoniumhalides.[182] From carbon-11 labeled guanidine and trimethinium salts as reagents several 2-^{11}C-2-aminopyrimidines are prepared by cyclocondensations.[181] Carbon-11 labeling of aliphatic and aromatic hydrocarbons was achieved by C-C bond formation using $^{11}CH_3I$. By this method, the Corey-House coupling 1-^{11}C-pentane, 1-^{11}C-nonane, 1-^{11}C-undecane and the methyl group containing aromatic hydrocarbons methyl-^{11}C-toluene and 2-^{11}C-methylnaphtalene were prepared.[183]

Several ^{11}C-labeled carboxylic acids were prepared by carboxylation of the appropriate grignard or organo-lithium compound. Machulla prepared ^{11}C-nicotinic, ^{11}C-anthranilic, and ^{11}C-salicylic acid.[14] The same author reports on the preparation of the amide from ^{11}C-nicotinic acid.[184] ^{11}C-acetic acid was prepared from methyl magnesium bromide.[104,185] On the synthesis of fatty acids also recent publications are available: octanoic acid,[186,187] palmitic acid,[104,187,188] and oleic acid.[104] Carbon-11 labeled alcohols were prepared from the corresponding carboxylic labeled acids by reduction with $LiAlH_4$. Good yields of methanol,[189] ethanol,[104,189,190] isopropanol,[189] and undecanol[104] were obtained. ^{11}C-alcohols can be converted conveniently into carbon-11 ethers.[191] ^{11}C-benzaldehyde and ^{11}C-octanal have been prepared from ^{11}C-labeled benzoic acid and octanoic acid, by oxidation of the intermediately prepared alcohols[192] or by insertion of ^{11}CO into the carbon-boron bond of an organoborane.[193]

Carboxylic labeled ^{11}C-DL-amino acids are prepared from $^{11}CO_2$ and α-lithiated isocyanides, by amination of α-bromo acids and by the Strecker synthesis. The first method is used to prepare ^{11}C-phenylglycine, ^{11}C-phenylalanine,[65] ^{11}C-labeled 3,4 dihydroxyphenylalanine,[194,195] and ^{11}C-alanine.[196] The last amino acid is also prepared from ^{11}C-propionic acid by bromination and subsequent amination.[14] By the Strecker synthesis, with H^{11}CN or

Na^{11}CN as precursor, ^{11}C-labeled DL-valine[197-199] and the aromatic amino acid DL-tryptophan[197,199,200-202] were prepared. The unnatural, alicyclic α-amino acids 1-aminocyclobutane carboxylic acid,[199,203] 1-aminocyclopentane carboxylic acid[87,199,204] and 1-aminocyclohexane carboxylic acid[199] were obtained by the same procedure.

The preparation of methyl labeled ^{11}C-L-methionine was achieved by the reaction of S-benzylhomocysteine[96] or L-homocysteine.[202,206] The methionine prepared was used as intermediate in ^{11}C-labeling of methionine-containing polypeptides.[207] From ^{11}CH$_3$I and an isocyanide Kloster prepared DL-3-^{11}C-alanine.[208] Partially resolved L-alanine labeled in the 3-position was obtained by an asymmetric synthesis.[196]

Examples of enzymatic amino acid preparation methods are found for L-^{11}C-alanine,[209] L-^{11}C-α-aminobutyric acid[210] and L-^{11}C-glutamic acid.[211] Depending upon the precursor used, ^{11}C-acetate or ^{11}C-carbon dioxide the α or ω position is labeled. The preparation of carbon-11 dialkyl, diaryl, and alkylaryl hydantoins by the modified Bücherer-Bergs reaction is described by Winstead[212] and Stavchansky.[213] The compounds were synthesized by heating H^{11}CN or K^{11}CN under pressure together with the corresponding aldehyde or ketone and an excess of (NH$_4$)$_2$CO$_3$. Roeda prepared an ^{11}C-alkylaryl- and a diarylhydantoin from ^{11}C-phosgene.[111] The same radioactive precursor was used for the preparation of ^{11}C-labeled urea,[214] ethyl chloroformate, diethylcarbonate, diphenylurea, 5,5-diethyl barbiturate, and 5,5-ethylphenyl barbiturate.[111] 1-^{11}C-hexobarbital was prepared by Kloster[215] by methylation of norhexobarbital with ^{11}CH$_3$I. When H^{11}CN is oxidized to ^{11}C-cyanate and added to hydroxylamine a mixture of hydroxyurea and isohydroxyurea is obtained.[216] Another application of ^{11}C-labeled hydrogen cyanide is in the preparation of nitrils, either for direct biomedical application or as intermediate in the preparation of ^{11}C-amines and carboxylic labeled acids. Fowler prepared 1-^{11}C-octane nitril and reduced this compound with LiAlH$_4$ to 1-^{11}C-octane amine.[217] This method was also used for ^{11}C-labeling of several other aliphatic amines[186] and aromatic amines.[218] ^{11}C-aliphatic diamines with 4 to 9 linear carbon atoms can be obtained by the reduction with BH$_3$ of ^{11}C-dinitriles, obtained by the reaction of the appropriate bromonitrile or dibromoalkane with H^{11}CN.[226] ^{11}C-labeled lactic acid was prepared through the intermediate ^{11}C-lactonitrile, obtained by addition of H^{11}CN to the bisulfite addition product of acetaldehyde.[219] The same method was used for the preparation of several aromatic nitrils and α-hydroxyacids. Besides ^{11}C-mandelonitrile the *p*-methoxy-, *p*-hydroxy and the 3,4-dihydroxy derivatives as well as m-methyl, *o*-chloro- and *p*-chloromandilonitrile were prepared as intermediates for the corresponding α-hydroxy acids and ethyl and benzyl esters.[220] Several ^{11}C-labeled 2-*N*-phenethyl- and benzyl-aminoalkanenitrils were prepared by reacting the respective aldehyde sodium bisulfite product with phenethylamine or benzylamine and Na^{11}CN[221-223] ^{11}C-labeled α-*p*-iodoanilinophenyl acetonitril was prepared from Na^{11}CN by addition of Schiff's base.[224] Carbon-11 labeled *N*-alkyl-*p*-iodobenzenesulfonamides are obtained via intermediate ^{11}C-cyanide followed by reduction to amines and condensation with *p*-iodobenzenesulfonylchloride.[225] The neurotransmitter dopamine and the halogenated analogue 6-iodopamine has been labeled with ^{11}C by Fowler.[218] Also on the labeling of the dimethoxy derivative of dopamine is reported.[227] The intermediate ^{11}C-nitril can be demethylated with BBr$_3$. ^{11}C-serotonin is prepared by essentially the same method. The only difference is that the nitril is reduced with LiAlH$_4$ and that the demethylation is carried out by hydrogenation using H$_2$ and Pd as catalyst.[228]

Two methods for the ^{11}C-methylation of amines are used in carbon-11 chemistry. At first the reductive methylation of free amino groups by H^{11}CHO according to the Leuckart reaction in the modification of Eschweiler-Clarke. Secondly the preparation of ^{11}C-labeled tertiary amines and quaternary ammonium salts using ^{11}CH$_3$I as radioactive precursor. The reductive methylation is used for the preparation of the ^{11}C-methylated-polyamines putrescine, spermine and spermidine,[229] the alkaloid ^{11}C-nicotine[230,231] and the psychoactive drugs ^{11}C-chlorpromazine, ^{11}C-thioproperazine. ^{11}C-imipramine, ^{11}C-chlorimipramine,[231,232] ^{11}C-clomipramine,[233,234] ^{11}C-etorphine, a compound for the study of brain opiate receptors[235] and O-methyl-bufotenine[236] were also obtained by reductive methylation.

The French group also was extensively involved in the preparation of ^{11}C-labeled secondary amines by [11]CH$_3$I methylation. They prepared ^{11}C-diazepam and ^{11}C-flunitrazepam for investigating benzodiazepine receptors.[237] The ^{11}C-labeling of secondary methylamines and the alkylation of amides was shown by the preparation N-^{11}C-methylbenzylamine and N-^{11}C-methylacetanilide.[104] For pharmacokinetic investigations Kloster used the same method for the labeling of ^{11}C-morphine which yields after acetylation ^{11}C-heroin.[238] ^{11}C-morphine can also be prepared by reductive alkylation with H[11]CHO and NaBH$_4$.[239] Other compounds prepared are ^{11}C-caffeine, ^{11}C-ephadrine, and ^{11}C-methylephedrine.[240] The synthesis of ^{11}C-labeled quarternary ammonium salts with [11]CH$_3$I is demonstrated in the preparation of 1,1'-dimethyl-4,4'-dipyridinium diiode by Palmer,[99] hexamethonium by Anwar,[241] ^{11}C-choline and ^{11}C-acetylcholine by Långström.[104] ^{11}C-phosgene was the radioactive precursor in the synthesis of ^{11}C-pimozide[242] a neuroleptic drug prepared to investigate the in vivo detection of dopamine receptors.

The in vivo investigation of estrogen receptors in hormone sensitive tumors is one of the aims for the preparation of ^{11}C-steroids. With ^{11}C-acetylene as precursor ^{11}C-labeled 17α-ethynylestradiol and its 17β-methoxy derivative moxestrol are prepared.[90,243] The addition of [11]CH$_3$Li to 17-keto steroids was used for the preparation of ^{11}C-labeled 17α-methylestradiol, 17α-methyl-testosterone, and 17α-methylestradiol 3-methyl ether.[244]

Nonradioactive methyl lithium forms with [11]CO$_2$ a complex that after hydrolysis yields ^{11}C-acetone.[245]

Interest in glucose metabolism has prompted several investigations to prepare ^{11}C-glucose and ^{11}C-glucose analogues. Carbon-11 labeled glucose, previously obtained by photosynthesis of Swiss chard and broad bean plants can also be prepared using algae.[246,247] The chemical preparation of 1-^{11}C-2-deoxy-D-glucose was accomplished by the Brookhaven group using Na[11]CN as precursor.[248,249] The 3-^{11}C-methyl-D-glucose analogue was prepared by methylation of the potassium salt of diacetone-D-glucose.[250,251] Besides the application of photosynthetic preparation methods, several enzymatic methods are recently used. Crawford reports on the enzymatic transfer of H[11]CHO to deoxyuridylate for the preparation of the deoxynucleotide.[252]

From this product ^{11}C-thymidine was obtained by enzymatic dephosphorylation.[252] This ^{11}C-nucleoside is also prepared organic synthetically by alkylation with [11]CH$_3$I of a lithiated nucleoside. Three different preparation methods are reported.[104]

Enzymatic methods are also used for the ^{11}C-labeling of citric acid,[253] pyruvic acid and L-lactic acid.[254] Starting with DL-3-^{11}C-alanine Kloster could prepare L-3-^{11}C-lactic acid, using a transaminase and lactate dehydrogenase.[208]

From ^{11}C-benzoic acid Gatley prepared biosynthetically ^{11}C-hippuric acid using mitochondria isolated from the rat liver.[255]

C. Nitrogen-13 Labeled Organic Compounds

After 1976 only a few nitrogen-13 labeled compounds are prepared. Most of the reported methods are on amino acids and start with [13]NH$_4^+$ which is enzymatically incorporated. The procedures mostly are improvements on procedures published earlier. New organic synthetic preparations are also scarce. The compound bis(2-chloroethyl)nitrosourea, an anticancer chemotherapeutic, has been labeled with ^{13}N in the nitroso group. As radioactive precursor [13]NO$_3^-$ produced by the (p,α) reaction on water, is used.[256,257] In an earlier report the same author described already a similar method, starting with ^{13}N-labeled ammonia.

The same chemical reaction was used in the preparation of ^{13}N-streptozotocin and ^{13}N-nitrosocarbaryl, also antitumor agents.[258] Krizek prepared ^{13}N-urea from [13]NH$_3$, silver cyanate and ammoniumchloride.[259]

As discussed before, an advantage of enzymatic methods is that very often the natural stereoisomers of the prepared compounds are obtained. A recent improvement in the preparation is the application of immobilized enzymes. Gelbard described the preparation of L-^{13}N-glutamate by this technique. The procedure to immobilize the enzyme used, is included.[260] Baumgartner developed a remote, semiautomated procedure not only for L-^{13}N-glutamate but also for L-glutamine, labeled either in the α or ω positron.[261] The semiautomated procedure for L-^{13}N-alanine is reported earlier by the same group.[262] Nitrogen-13 labeled L-glutamic acid is used as intermediate in the preparation of two aromatic amino acids. L-^{13}N-tyrosine and L-^{13}N-phenylalanine are prepared by transamination of the appropriate α-keto acid with glutamate oxalacetate transaminase isolated from a pig heart.[263] The enzymatic synthesis of ^{13}N-asparagine is described in an abstract by Majumdar.[134] A new nonenzymatic preparation is reported by Elmaleh.[264] The β-carboxylic group of the protected L-aspartic acid is activated with N-hydroxysuccinimide. This intermediate compound is refluxed with ^{13}N-ammonia, hydrolyzed and after chromatography L-^{13}N-asparagine is obtained in good yields.

VI. GENERATORS FOR SHORT-LIVED POSITRON EMITTERS

A. The ^{52}Fe-^{52}ΣMn System

Iron-52 decays with a half-life of 8.3 hr by positron emission (57.8%) and by electron capture to the 546 keV level in manganese-52.[261] De-excitation of the manganese-52 nucleus occurs via emission of a 168.8 keV gamma ray to a metastable level in ^{52}Mn. ^{52}Mn decays by positron emission (97%) with a half-life of 21.3 m. The maximum positron energy is 2.63 MeV and the range corresponding to this positron energy in water is 13 mm. In addition to the annihilation radiation a γ-ray of 1434 keV is emitted. In 2% of the cases ^{52}ΣMn is decaying via an isomeric transition to ^{52}Mn, which has a half-life of 5.67 d. Iron-52 can be obtained via a helium-3 or helium-4-induced reaction on a chromium target, by proton bombardment of a manganese target or by proton-induced spallation reactions on a nickel target.[267-270] The manganese-52m can be obtained by eluting an anion-exchange resin column with HCl.[271] The short half-life of ^{52}ΣMn is very well suited for use in sequential studies. Animal studies indicate that ^{52}ΣMn, due to the rapid clearance of the blood and the adequate myocardial concentration, is a good agent for myocardial imaging and is comparable to the widely used ^{201}Tl.[271-274]

B. The ^{62}Zn-^{62}Cu System

Zinc-62 decays with a half-life of 9.13 h by positron emission (8.4%) and electron capture to ^{62}Cu. The half-life of copper-62 is 9.74 m and decays by positron emission (97.5%) and electron capture to stable nickel-62. The most prominent γ-ray associated with the decay of copper-62, with an abundance of 0.335%, has an energy of 1173 keV.[275] The maximum positron energy is 2.93 MeV and the corresponding range of positrons of this energy in water is 14.5 mm. Zinc-62 can be produced by irradiating a copper foil with protons.[276,277] The copper is obtained by eluting the generator with 3 mℓ of 0.1 N HCl containing 100 mg NaCl and 1 μg Cu(II) carrier per mℓ.[278] A number of ^{62}Cu-labeled radiopharmaceuticals like CuS, CuO, Cu-EDTA and Cu-DTPA have been prepared.[276,278]

C. The ^{68}Ge-^{68}Ga System

Germanium-68 with a half-life of 287 d decays to gallium-68. The half-life of gallium-68 is 68.3 m and decays in 90% of the cases by positron emission. The only γ-ray in the decay of ^{68}Ga with an abundance of more than 1% is a γ-ray of 1.077 MeV (3.2%).[279] The maximum positron energy is 1.9 MeV and the corresponding range of positrons of this energy in water is 9 mm. ^{68}Ge can be produced via the ^{69}Ga(p,2n)^{68}Ge reaction. The generator is commercially available and based on the work of Greene and Tucker.[280] The ^{68}Ga is eluted with a solution of ethylenediaminetetraacetic acid (EDTA). Since ^{68}Ga-EDTA has a very high stability constant it is hard to decompose this chelate for the preparation of other radiopharmaceuticals. Welch and Wagner have given an overview[281] of the radiopharmaceuticals which have been prepared after decomposition of the ^{68}Ga-EDTA chelate.

The use of ^{68}Ga would greatly be facilitated by the availability of ionic gallium-68. For this reason several attempts[282-288] have been made to replace the EDTA as eluant. ^{68}Ga yields between 49% and 90% have been reported in literature.[282,284] Recently ^{68}Ga-labeled dihydroxyanthraquinones have been used for the imaging of kidney and liver.[289] When gallium-68 becomes available commercially in an ionic form this radionuclide may become of great importance: its position for positron emission tomography would be the same as the position of technetium-99m for the conventional scintigraphy.

D. The ^{82}Sr-^{82}Rb System

Strontium-82 decays with a half-life of 25 d by electron capture to the ground state of rubidium-82. The half-life of ^{82}Rb is 75 s and decays for 95.3% by positron emission to stable krypton-82. The only γ-ray with an abundance of more than 1% present in the decay of ^{82}Rb is a γ-ray with an energy of 776.5 keV and an abundance of 13.4%.[290] The maximum positron energy is 3.36 MeV and the corresponding range of positrons of this energy in water is nearly 17 mm. The strontium-82 can be produced via the ^{85}Rb(p,4n)^{82}Sr reaction or by proton induced spallation reactions on molybdenum.[291] Both cation[292-294] and chelating[295] ion exchange resins have been used in this generator system. An evaluation of rubidium-82 generators is given by Yano and Budinger.[293] A good localization of the rubidium-82 is achieved in myocardial and kidney tissue.[291]

E. The ^{122}Xe-^{122}I System

Xenon-122 decays with a half-life of 20.1 hr by electron capture to iodine-122. ^{122}I decays, with a half-life of 3.62 m by positron emission (76%) and by electron capture to stable ^{122}Te.[296] The most abundant γ-ray (17.7%) has an energy of 546 keV. The maximum positron energy is 3.1 MeV and the corresponding range of positrons of this energy in water is 15 mm. The xenon-122 can be produced via the proton-induced reaction on iodine-127. It can be obtained as a by-product in the production of ^{123}Xe if the available proton energy is sufficient.[297] This iodine nuclide is, due to its short half-life, the best of the iodine nuclides when a number of successive rapid diagnostic procedures have to be performed.

VII. SOME OTHER SHORT-LIVED POSITRON-EMITTING NUCLIDES

One glance at the chart of nuclides and it is obvious that the number of positron-emitting nuclides discovered up till now is very large. Most of these nuclides have been produced and studied for scientific reasons. A selected number, depending on the half-life and the radiation characteristics, are produced for mainly medical or biological research. The production and possible use of a few will be described below shortly.

A. Neon-19

The half-life of neon-19 is 17.3 s. This nucleus decays by positron emission (99%) and electron capture to stable fluorine-19.[298] The maximum positron energy is 2.24 MeV and the corresponding range of positrons of this energy in water is nearly 11 mm. This high positron energy can affect the spatial resolution obtainable, depending on the positron imaging device used. Neon-19 can be produced via the ^{19}F(p,n)^{19}Ne and the ^{16}O(α,n)^{19}Ne reaction. The yield of the latter is 1.2 mCi/μA.m when a 23 MeV beam is used.[299] The neon is produced for lung ventilation studies. Due to its half-life this radionuclide can be regarded as the positron-emitting equivalent of ^{81}ΣKr. The solubility of neon in water is about the

Table 5
SOLUBILITY OF SOME GASES IN WATER[300]

Gas	Isotope	Solubility (cm³/100 g H$_2$O)	T (°C)
Nitrogen	^{13}N	1.42	40
Neon	^{19}Ne	1.47	20
Krypton	^{81m}Kr	6.0	25
Xenon	^{133}Xe	11.9	25

Table 6
RELEVANT REACTIONS FOR THE PRODUCTION OF ^{30}P

Particle	Reaction	Q-value (MeV)	E(thres.) (MeV)	σ(th.)[a] (mb)	σ(exp) (mb)	Ref.
p	^{31}P(p,pn)^{30}P	−12.3	12.7	80	250	302
d	^{32}S(d,α)^{30}P	4.9	0	250		
^{3}He	^{28}Si(^{3}He,p)^{30}P	6.4	0	130		
^{4}He	^{27}Al(^{4}He,n)^{30}P	−2.6	3.0	240	400	302
	^{28}Si(^{4}He,pn)^{30}P	−14.2	16.2	130		

[a] Cross-section calculated according to the statistical model.

same as for nitrogen but much lower than for the commonly used xenon-133, see Table 5. The solubility is of importance because it defines part of the background in a ventilation study.

B. Phosphorus-30

The half-life of phosphorus-30 is 2.5 m. This nuclide decays by positron emission to stable silicon-30.[301] The maximum positron energy is 3.25 MeV and the corresponding range of positrons of this energy in water is 16 mm. The high maximum positron energy may be a hindrance to the spatial resolution obtainable, depending on the imaging device used. Although ^{30}P was already discovered in 1934 by Curie and Joliot excitation functions were only recently published. There are five relevant reactions for the production of phosphorus-30 with charged particles as shown in Table 6. The excitation functions for two of these reactions have been measured.[302] The α-particle induced reaction on ^{27}Al was preferred above the proton-induced reaction on phosphorus for two reasons: firstly the proton-induced reaction inhibits high specific activities and secondly because of the easy target processing when aluminum is used. The yield for the α-particle induced reaction is 4.7 mCi/μA m when 28 MeV beam and an 18 MeV thick target is used. The deuteron-induced reaction on sulfur-32 has not been measured but the positive Q-value and the high theoretical cross-section suggest some possibilities. The advantage of phosphorus-30 above the in life sciences most commonly used phosphorus-31 and 32 is the possibility of the easy external detection by measuring the annihilation radiation instead of using β-detectors.

C. Potassium-38

The half-life of potassium-38 is 7.6 m. This nucleus decays by pure positron emission to stable argon-38. In 99.8% of the cases ^{38}K decays to the 2.168 MeV level in ^{38}Ar and 0.2% of the cases to the 3.94 MeV level.[303] The subsequent γ-rays from the de-excitation of the argon-38 nucleus have energies of 2.168 MeV (99.8%) and 3.936 MeV (0.2%). The maximum positron energy is 2.68 MeV and the corresponding range of positrons of this energy

Table 7
PRODUCTION RATES OF ^{38}K

Target	Reaction	Energy (MeV)	Production rate (mCi/μA)	Ref.
Ar	^{40}Ar(p,3n)^{38}K	32	5.2	304
CaO	^{40}Ca(d,^{4}He)^{38}K	5.5	—	305
KCl	^{39}K(^{3}He,^{4}He)^{38}K	7.8	0.28	306
	^{37}Cl(^{3}He,2n)^{38}K	23	0.12	306
NaCl	^{35}Cl(^{4}He,n)^{38}K	11	—	305
CCl$_4$	^{35}Cl(^{4}He,n)^{38}K	14.7	0.6	306

Table 8
PRODUCTION RATES OF ^{52}Fe

Target	Reaction	Energy (MeV)	Production rate (mCi/μA)	Ref.
Mn	^{55}Mn(p,4n)^{52}Fe	65	2	268
	^{55}Mn(p,4n)^{52}Fe	70	2.3	270
Cr	^{52}Cr(^{3}He,3n)^{52}Fe	23	0.01	269
	^{52}Cr(^{3}He,2n)^{52}Fe	40	0.65	18
	^{52}Cr(^{3}He,3n)^{52}Fe	46.5	0.6	267
	^{50}Cr(^{4}He,2n)^{52}Fe	30	0.02	310
	^{50}Cr(^{4}He,2n)^{52}Fe	40	0.1	312
	^{50}Cr(^{4}He,2n)^{52}Fe	65	0.02	311

in water is 13 mm. This high maximum positron energy will have influence on the spatial resolution obtainable.

In Table 7 the different production reactions used and their yields are listed. From this table it is obvious that the proton-induced reaction is the best for the production of ^{38}K. The deuteron, helium-3, and helium-4-induced reactions have the advantage that only low beam energies, available on a compact medical cyclotron, are required. The separation of ^{38}K from the target materials are also reported in the references of Table 7. The interest of potassium is in its application for myocardial imaging[307] and for rapid turn-over studies.[308]

D. Iron-52

The half-life of iron-52 is 8.27 h. This nuclide decays by positron emission (56.5%) and electron capture to the 546 keV level in manganese-52.[309] The de-excitation of the manganese nucleus occurs via emission of a 168.8 keV γ-ray to the metastable level, half-life 21.3 m, at 377.8 keV. ^{52}ΣMn decays in 98% of the cases to stable chromium-52, and for 2% via an isomeric transition to manganese-52 with a half-life of 5.67 d. The maximum positron energy is 0.804 MeV and the corresponding range of positrons of this energy in water is 5 mm. In Table 8 the different production reactions used and their yields are listed. From this table it is obvious that the proton reaction on manganese is the best way to produce iron-52. The high energy, 65 MeV, of the proton beam inhibits a wide spread use of this method. In our laboratory this reaction is used. A two hour chemical processing of the metallic manganese target is required to obtain the iron-52. The production rate measured is the same as reported by Saha and Farrer.[268] If a high energy, 40 MeV, helium-3 beam is available a reasonable amount of iron-52 can be made by bombarding a chromium target. The yield of iron-52 by bombarding chromium with helium-4 particles has a low yield due to the low abundance of chromium-50, 4.35%. Historically this was the first reaction used for the production of iron-52 as reported by Yano and Anger.[311] Murakami and co-workers[312] compared the helium-3 and helium-4-induced reactions. They favored the helium-3-induced reaction not only because of the higher yield but also because of the much higher radionuclidic purity. The impurity of iron-55 is 6.3% for the helium-4-induced reaction and 0.07% for the helium-3-induced reaction. Iron-52 can also be produced by spallation reactions. Trial production runs with 200 and 800 MeV proton beams on a nickel target show, as reported by Atcher and co-workers,[270] a high yield. Iron-52 is used for the scintigraphy of bone marrow.[313] The erythropoietic function may be studied with iron-59 but the radiation characteristics inhibit an acceptable image.

E. Rubidium-81

The half-life of rubidium-81 is 4.58 h. This nuclide decays by positron emission (73%) and electron capture to krypton-81.[314] Only 2.8% of the decay is direct to the ground state of krypton-81, half-life 2.1×10^5 y. In all other cases the decay goes via the metastable level at 190 keV in the krypton-81 nucleus. The half-life of $^{81}\Sigma Kr$ is 13 s. The decay of ^{81}Rb is mainly characterized by annihilation radiation and gamma radiation of 190 keV (66%) and 446 keV (19%).[314] The maximum positron energy is 1.05 MeV and the corresponding range of positrons of this energy in water is 4.5 mm. The production reaction to be used will depend on the application of the rubidium-81. If it has to be used in a $^{81}\Sigma Kr$ generator system the radionuclidic purity is not important since $^{82}\Sigma Rb$, ^{82}Rb, ^{83}Rb, and ^{84}Rb all decay to stable krypton nuclides. If the ^{81}Rb has to be injected and both species, the ^{81}Rb and the ^{81}Kr nuclei, are used to obtain diagnostic information as suggested by Jones and Matthews,[315] the radionuclidic purity is of importance. In this part only those reactions yielding rubidium-81 with a high radionuclidic purity will be considered. ^{81}Rb can be produced directly by bombarding krypton gas, preferable isotopically enriched. Isotopically enriched krypton up to 90% is commercially available but rather expensive. All nuclear reactions with more than one outgoing particle, like the $^{82}Kr(p,2n)^{81}Rb$ reaction, will yield a contamination with ^{82}Rb. Since the half-life of ^{82m}Rb is 6.3 hr its disturbing influence will always be present. The presence of ^{82m}Rb is due to pre-equilibrium effects which are of importance at energies well above the threshold energy of the $^{82}Kr(p,n)^{82m}Rb$ reaction. So the only reaction left to produce pure ^{81}Rb from a krypton nuclide is the $^{80}Kr(d,n)^{81}Rb$ reaction. This reaction is used by Harper and co-workers[316] with a 7.4 MeV deuteron beam. Due to the fact that the krypton gas used consisted for only 37.2% out of ^{82}Kr also ^{79}Rb, ^{83}Rb, and ^{84}Rb were produced. No ^{82}Rb was observed due to the fact that the cross-section of the $^{80}Kr(d,\gamma)^{82m}Rb$ reaction is very small. However, by waiting a few hours the main contaminant ^{79}Rb has removed itself. Five hours after the end of bombardment the yield of ^{81}Rb was 350 µCi/µAh with 0.46 µCi/µAh of ^{83}Rb and 0.07 µCi/µAh of ^{84}Rb.[316] All helium-3 or helium-4-induced reactions on natural bromium or enriched bromium-79 targets yield a product at least contaminated with ^{82m}Rb. Even the helium-3-induced reaction on ^{79}Br showed a substantial amount of ^{82m}Rb due to the $^{79}Br(^3He,\gamma)^{82m}Rb$ reaction.[317]

The procedure for the production of pure ^{81}Rb is via an indirect method comparable to the production of pure iodine-123. The mother nuclide of rubidium-81, strontium-81, can be produced via the $^{85}Rb(p,5n)^{81}Sr$ or the $^{85}Rb(\alpha,p7n)^{81}Sr$ reaction[318-322] at, respectively, 70 MeV and 135 MeV. After the irradiation the produced strontium nuclides are separated from the target material and via the $^{85}Rb(p,pxn)^{85-x}Rb$ produced nuclides. After the optimum waiting time of 96.4 m, calculated from the half-lives of ^{81}Sr and ^{81}Rb, a second separation is carried out, yielding the ^{81}Rb formed by the decay of $^{81}Sr(T1/2 = 26$ m$)$. Due to the half-life of the also produced $^{82}Sr(T1/2 = 25$ d$)$ the amount of ^{82m}Rb in the final product can be kept below 0.1%. The yields of pure ^{81}Rb are summarized in Table 9. If the injected ^{81}Rb-^{81m}Kr only has to be followed in time in a very restricted area in the human body a Ge(Li) detector would allow for even a large amount of contamination with ^{82m}Rb due to

Table 9
YIELDS OF PURE ^{81}Rb

Target	Reaction	Energy (MeV)	Yield	Impurities	Ref.
Kr	^{80}Kr(d,n)^{81}Rb	7.4	350 μCi/μAh[a]	0.13% ^{83}Rb	316
RbCl	^{85}Rb(p,5n)^{81}Sr	69	200 μCi/μA. gcm^{-2} [b]	—	318
RbCl	^{85}Rb(p,5n)^{81}Sr	70	150 μCi/μA.15 mg[c]	<1% ^{82m}Rb	319
RbCl	^{85}Rb(p,5n)^{81}Sr	65.5	600 μCi/μA.49 m[c]	<0.01% ^{82m}Rb	320
^{85}RbCl	^{85}Rb(p,5n)^{81}Sr	70	2.3 mCi/μAh[c]	—	321
RbCl	^{85}Rb(p,5n)^{81}Sr	65	830 μCi/μAh[c]	<0.1% ^{82m}Rb	322
RbCl	^{85}Rb(α,p7n)^{81}Sr	135	150 μCi/μAh[c]	<0.1% ^{82}Rb	322

[a] 5 h after EOB.
[b] 21 h after EOB.
[c] 96.4 m after the first separation.

the high energy resolution of this type of detector.[323] Harper and co-workers[316] have used ^{81}Rb-^{81m}Kr in heart and kidney studies but concluded, in 1974, that a conventional scintillation camera does not have the precision needed for clinical studies. Iodine and co-workers[319] used pure ^{81}Rb-^{81m}Kr as well as contaminated ^{81}Rb-^{81m}Kr, up to 25% ^{82m}Rb, in a dynamic phantom. This study suggests that ^{81}Rb-^{81m}Kr can be used for regional blood flow measurements in vivo with currently available NaI detection systems after the injection of ^{81}Rb into the arterial systems supplying the organ of interest. Because they did not use the ^{81m}Kr-^{81}Rb ratio, also contaminated ^{81}Rb could be used. Lambrecht and co-workers[324] used a mixture of ^{81}Rb and ^{82m}Rb for myocardial studies in animals with a quantitative positron emission transaxial tomograph and a positron camera. Quantitative reproducible images of the myocardium were obtained.

REFERENCES

1. **Evans, R. D.**, The neutrino, in *The Atomic Nucleus,* McGraw-Hill, New York, 1969, 541.
2. **Auble, R. L.**, Nuclear Data Sheets for A = 123, Nuclear Data Sheets B7, 363, 1972.
3. **Konopinsky, E. J. and Rose, M. E.**, The theory of nuclear β-decay, in *Alpha- Beta- and Gamma-Ray Spectroscopy,* Vol. 2, Siegbahn, K., Ed., North-Holland, Amsterdam, 1965, 1340.
4. **Enge, H. A.**, *Introduction to Nuclear Physics,* Addison-Wesley, Reading, 1969, 482.
5. **Marmier, P. and Sheldon, E.**, *Physics of Nuclei and Particles,* Vol. 1, Academic Press, New York, 1969, 146.
6. **Deutsch, M. and Berko, S.**, Positron annihilation and positronium, in *Alpha- Beta- and Gamma-Ray Spectroscopy,* Vol. 2, Siegbahn, K., Ed., North-Holland, Amsterdam, 1969, chap. 26.
7. **DeBenedetti, S., Cowan, C. E., Kronneker, W. R., and Primakoff, H.**, On the angular distribution of two photon annihilation radiation, *Phys. Rev.,* 77, 205, 1950.
8. **Wolf, A. P. and Redvanly, C. S.**, Carbon-11 and radiopharmaceuticals, *Int. J. Appl. Radiat. Isotopes,* 28, 29, 1977.
9. **Straatmann, M. G.**, A look at ^{13}N and ^{15}O in radiopharmaceuticals, *Int. J. Appl. Radiat. Isotopes,* 28, 13, 1977.
10. **Wolf, A. P.**, Synthesis of organic compounds labeled with positron emitters and the carrier problem, *J. Label. Comp. Radiopharm.,* 18, 1, 1981.
11. **Wolf, A. P.**, *Advances in Physical Organic Chemistry,* Vol. 2, Gold, V., Ed., Academic Press, New York, 1964, 201.
12. **Stöcklin, G.**, *Chemie heisser Atome,* Verlag Chemie., Weinheim BRD., 1969.
13. **Långström, B. and Bergson, G.**, The determination of optimal yields and reaction times in syntheses with short-lived radionuclides of high specific radioactivity, *Radiochem. Radioanal. Lett.,* in press.

14. **Machulla, H. J., Laufer, P., and Stöcklin, G.,** Radioanalytical quality control of ^{11}C, ^{18}F, and ^{123}I-labelled compounds and radiopharmaceuticals, *J. Radioanal. Chem.*, 32, 381, 1976.
15. **Walsh, W. F., Fill, H. R., and Harper, P. V.,** Nitrogen-13-labelled ammonia for myocardial imaging, *Sem. Nucl. Med.*, 7, 59, 1977.
16. **Rankowitz, S., Robertson, J. S., Higinbotham, W. A., and Rosenblum, M. J.,** Positron scanner for locating brain tumours, *IRE Intern. Convention Record*, 10, 49, 1962.
17. **Yamamoto, Y. L., Thompson, C. J., Meyer, E., Robertson, J. S., and Feindel, W.,** Dynamic positron emission tomography for the study of cerebral dynamics in a cross section of head using positron emitting ^{68}Ga-EDTA and ^{77}Kr, *J. Comp. Asst. Tomog.*, 1, 43, 1977.
18. **Ter-Pogossian, M. M., Phelps, M. E., Hoffman, E. J., and Mullani, N. A.,** A positron emission transaxial tomograph for nuclear imaging (PETT), *Radiology*, 114, 89, 1975.
19. **Phelps, M. E., Hoffman, E. J., Mullani, N. A., and Ter-Pogossian, M. M.,** *J. Nucl. Med.*, 16, 210, 1975.
20. **Phelps, M. E., Hoffman, E. J., Sung-Cheng Huan, and Kuhl, D. E.,** ECAT: a new computerized tomographic imaging system for positron emitting radiopharmaceuticals, *30Nucl. Med.*, 19, 635, 1978.
21. **Cho, Z. H., Cohen, M. B., Singh, M., Erikson, L., Chan, J., and MacDonald, N.,** Performance and evaluation of the circular ring transverse axial positron camera (CRTAPC), *IEEE Trans. Nucl. Sci.*, NS-24, 532, 1977.
22. **Bohm, Chr., Erikson, L., Bergstrom, M., Litton, J., Sundman, R., and Singh, M.,** A computer assisted ring detector positron camera system for reconstruction tomography of the brain, *IEEE Trans. Nucl. Sci.*, NS-25, 624, 1978.
23. **Derenzo, S. E. and Vuletich, T.,** The Donner 280-crystal high resolution positron tomograph, *IEEE Trans. Nucl. Sci.*, NS-26, 2790, 1979.
24. **Brownell, G. L., Correia, J. A., and Zamenhof, R. G.,** Positron instrumentation, in *Recent Advances in Nuclear Medicine*, Vol. 5, Lawrence, J. H. and Budinger, T. F., Eds., Grune & Stratton, New York, 1978, 1.
25. **Brownell, G. L., Burnham, C. A., Chesler, D. A., Correia, J. A., Correll, J. E., Hoop, B., Jr., Parker, J. A., and Subramanyam, R.,** Transverse section imaging of radionuclide distributions in the heart, lung, and brain, in *Reconstruction Tomography in Diagnostic Radiology and Nuclear Medicine*, Ter-Pogossian, M. M., Phelps, M. E., Brownell, G. L., Cox, J. R., Jr., Davis, D. O., and Evens, R. G., Eds., University Park Press, Baltimore, 1977, 293.
26. **Burnham, C. A. and Brownell, G. L.,** A multicrystal positron camera, *IEEE Trans. Nucl. Sci.*, NS-19, 201, 1972.
27. **Brownell, G. L., Burnham, C. A., Correia, J., Chesler, D., Ackerman, R., and Tavares, J.,** Transverse section imaging with the MGH positron camera, *IEEE Trans. Nucl. Sci.*, NS-26, 2698, 1979.
28. **Muehllehner, G., Buchin, M. P., and Dudek, J. H.,** Performance parameters of a positron imaging camera, *IEEE Trans. Nucl. Sci.*, NS-23, 528, 1976.
29. **Muehllehner, G., Atkins, F., and Harper, P. V.,** Positron camera with longitudinal and transverse tomographic ability, in *Medical Radionuclide Imaging*, Vol. 1, International Atomic Energy Agency, Vienna, 1977, 291.
30. **Anger, H. O. and Rosenthal, D. J.,** Scintillation camera and positron camera, in *Medical Radioisotope Scanning*, IAEA and WHO, Vienna, 1959, 176.
31. **Paans, A. M. J., de Graaf, E. J., Welleweerd, J., Vaalburg, W., and Woldring, M. G.,** Performance parameters of a longitudinal tomographic positron imaging system, *Nucl. Instr. Meth.*, 192, 491, 1982.
32. **Perez-Mendez, V., Kaufman, L., Lim, C. B., Price, D. C., Blumin, L., and Cavalieri, R.,** Multiwire proportional chambers in nuclear medicine present status and perspectives, *Int. J. Nucl. Med. Biol.*, 3, 29, 1976.
33. **Jeavons, A., Kull, K., Lindberg, B., Lee, G., Townsend, D., Frey, P., and Donath, A.,** Proportional chamber positron camera for medical imaging, *Nucl. Instr. Meth.*, 176, 89, 1980.
34. **Jeavons, A., Schorr, B., Kull, K., Townsend, D., Frey, P., and Donath, A.,** Large area stationary positron camera using wire chambers, in *Medical Radionuclide Imaging*, International Atomic Energy Agency, Vienna, 1980, in press.
35. **Mullani, N. A., Higgings, C. S., Hood, J. T., and Currie, C. M.,** Pett IV: design analysis and performance characteristics, *IEEE Trans. Nucl. Sci.*, NS-25, 180, 1978.
36. **Mullani, N. A., Ter-Pogossian, M. M., Carols, S., Higgins, C. S., Hood, J. T., and Ficke, D. C.,** Engineering aspects of PETT V, *IEEE Trans. Nucl. Sci.*, NS-26, 2703, 1979.
37. **Chu, G. and Tam, K. C.,** Three dimensional imaging in the positron camera using Fourier techniques, *Phys. Med. Biol.*, 22, 245, 1977.
38. **Townsend, D., Schorr, B., and Jeavons, A.,** Three dimensional image reconstruction for a positron camera with limited angular acceptance, *IEEE Trans. Nucl. Sci.*, NS-27, 463, 1980.
39. **Budinger, T. F., Gullberg, O. T., and Huesman, R. H.,** Emission computed tomography, in *Image Reconstructions from Projections*, Herman, G. T., Ed., Springer Verlag, New York, 1979, 147.

40. **Lederer, C. M. and Shirley, V. S.**, *Table of Isotopes,* 7th ed., John Wiley & Sons, New York, 1978, 8.
41. **Weisskopf, V. F. and Ewing, D. H.**, On the yield of nuclear reactions with heavy elements, *Phys. Rev.,* 57, 472, 1940.
42. **Blann, M. and Bisplinghoff, J.**, Hybrid Code Description, Rochester Report 3494-27, 1975.
43. **Hylten, G.**, An investigation of the $^{12}C(\gamma,n)^{11}C$ reaction for Eγ up to 1 GeV, *Nucl. Phys.,* A158, 225, 1970.
44. **Wieland, B. W., Highfill, R. R., and King, P. H.**, Proton accelerator targets for the production of ^{11}C, ^{13}N, ^{15}O, and ^{18}F, *IEEE Trans. Nucl. Sci.,* NS-26, 1713, 1979.
45. **Hintz, N. M. and Ramsey, N. F.**, Excitation functions to 100 MeV, *Phys. Rev.,* 88, 19, 1952.
46. **Valentin, L.**, Reaction (p,n) et (p,pn) Induites a Moyenne Energie sur des Noyaux Legers, *Nucl. Phys.,* 62, 81, 1965.
47. **Aamodt, R. L., Peterson, V., and Phillips, R.**, $^{12}C(p,pn)^{11}C$ cross section from threshold to 340 MeV, *Phys. Rev.,* 88, 739, 1952.
48. **Panofsky, W. K. H. and Phillips, R.**, Evidence for a (p,d) reaction in carbon, *Phys. Rev.,* 74, 1732, 1948.
49. **Cumming, J. B.**, Absolute cross section for the $^{12}C(p,pn)^{11}C$ reaction at 50 MeV, *Nucl. Phys.,* 49, 417, 1963.
50. **Measday, D. F.**, The $^{12}C(p,pn)^{11}C$ reaction from 50 to 160 MeV, *Nucl. Phys.,* 78, 476, 1966.
51. **Whitehead, A. B. and Foster, J. S.**, Activation cross sections for $^{12}C(p,pn)^{11}C$, $^{16}O(p,\alpha)^{13}N$ and $^{19}F(p,pn)^{18}F$, *Can. J. Phys.,* 36, 1276, 1958.
52. **Epherre, E. M. and Seide, C.**, Excitation functions of ^{7}Be and ^{11}C produced in nitrogen by low-energy protons, *Phys. Rev.,* C3, 2167, 1971.
53. **Nozaki, T., Ukuo, T., Akutsu, H., and Furukuwa, M.**, The radioactivation analysis of semiconductor graphite for nitrogen by the $^{14}N(p,\alpha)^{11}C$ reaction, *Bull. Chem. Soc. Jpn.,* 39, 2685, 1966.
54. **Brill, O. D. and Sumin, L. V.**, *Soviet J. At. Energy* (English Translation), 7, 856, 1960.
55. **Wohlleben, K. and Schuster, E.**, Der totale Wirkungsquerschnitt der Reaktionen $^{10}B(d,n)^{11}C$, $^{14}N(d,n)^{15}O$ und $^{16}O(d,n)^{17}F$ bis 3.2 MeV, *Radiochim. Acta,* 12, 75, 1969.
56. **Wilkinson, D. H.**, Reactions $^{12}C(d,n)^{13}N$ groundstate and $^{12}C(d,t)^{11}C$ up to Ed = 20 MeV, *Phys. Rev.,* 100, 32, 1955.
57. **Crandall, W., Millburn, P., Pyle, R., and Birnbaum, W.**, $^{12}C(x,xn)^{11}C$ and $^{27}Al(x,xp2n)^{24}Na$ cross sections at high energies, *Phys. Rev.,* 101, 329, 1956.
58. **Brill, O. D.**, ^{3}He-light nucleus interaction cross sections, *Sovjet J. Nucl. Phys.* (English Translation), 1, 37, 1965.
59. **Hahn, R. L. and Ricci, E.**, Interaction of ^{3}He particles with ^{9}Be, ^{12}C, ^{16}O and ^{19}F, *Phys. Rev.,* 146, 650, 1966.
60. **Lindner, M. and Osborne, R. N.**, Energy dependence of the cross section for the reaction $^{12}C(\alpha,\alpha n)^{11}C$, *Phys. Rev.,* 91, 1501, 1953.
61. **Christman, D. R., Crawford, E. J., Friedkin, M., and Wolf, A. P.**, Detection of DNA synthesis in intact organisms with positron-emitting (methyl-^{11}C) Thymidine, *Proc. Natl. Acad. Sci. USA,* 69, 1972, 988.
62. **Christman, D. R., Finn, R. D., Karlstrom, K. I., and Wolf, A. P.**, The production of ultra high activity ^{11}C-labelled hydrogen cyanide carbon dioxide, carbon monoxide and methane via the $^{14}N(p,\alpha)^{11}C$ reaction, *Int. J. Appl. Radiat. Isotopes,* 26, 435, 1975.
63. **Machulla, H. J., Laufer, P., and Stöcklin, G.**, ^{11}C-labelled radiopharmaceuticals: synthesis and high pressure liquid chromatography of nicotine-^{11}C acid, *Radiochem. Radioanal. Lett.,* 18, 275, 1974.
64. **Clark, J. C. and Buckingham, P. D.**, *Short-Lived Radioactive Gases for Clinical Use,* Butterworths, London, 1975, chap. 7.
65. **Vaalburg, W., Beerling-van der Molen, H. D., Reiffers, S., Rijskamp, A., Woldring, M. G., and Wynberg, H.**, Preparation of carbon-11 labelled phenylalanine and phenylglycine by a new amino acid synthesis, *Int. J. Appl. Radiat. Isotopes,* 27, 153, 1976.
66. **Ache, H. J. and Wolf, A. P.**, The effect of radiation on the reactions of recoil carbon-11 in the nitrogen-oxygen system, *J. Phys. Chem.,* 72, 1988, 1968.
67. **Vaalburg, W.**, Unpublished results.
68. **Wolf, A. P.**, Synthesis of organic compounds labelled with positron emitters and the carrier problem, *J. Label. Comp. Radiopharm.,* 18, 1, 1981.
69. **Berger, B., Mazière, M., Sastre, J., and Comar, D.**, Carrier-free ^{11}C-formaldehyde, an approach, *J. Label. Comp. Radiopharm.,* 17, 59, 1980.
70. **Crane, H. R. and Lauritsen, C. C.**, Further experiments with artificially produced radioactive substances, *Phys. Rev.,* 45, 497, 1934.
71. **Kamen, M. D.**, *Radioactive Tracers in Biology,* Academic Press, New York, 1947, 149.
72. **Buckingham, P. D. and Forse, G. R.**, The preparation and processing of radioactive gases for clinical use, *Int. J. Appl. Radiat. Isotopes,* 14, 439, 1963.

73. **Welch, M. J. and Ter-Pogossian, M. M.**, Preparation of short half-lived radioactive gases for medical studies, *Radiat. Res.*, 36, 580, 1968.
74. **Clark, J. C. and Buckingham, P. D.**, The preparation and storage of carbon-11 labelled gas for clinical use, *Int. J. Appl. Radiat. Isotopes*, 22, 639, 1971.
75. **Winstead, M. B., Lamb, J. F., and Winchell, H. S.**, Relationship of chemical structure to in vivo scintigraphic distribution patterns of ^{11}C compounds: I,^{11}C-carboxylates, *J. Nucl. Med.*, 14, 747, 1973.
76. **Ritchie, A. I. M.**, The production of the radioisotopes ^{11}C, ^{13}N and ^{15}O using the deuteron beam from a 3 MeV van de Graaff accelerator, *Nucl. Instr. Meth.*, 64, 181, 1968.
77. **More, R. D. and Troughton, J. H.**, Production of ^{11}C with a 3-MeV van de Graaff accelerator, *Int. J. Appl. Radiat. Isotopes*, 23, 344, 1972.
78. **Perris, A. G., Lane, R. O., Tong, J. Y., and Matthews, J. D.**, The production of carbon-11 for medical uses by a tandem van de Graaff accelerator by the reaction $^{11}B(p,n)^{11}C$, *Int. J. Appl. Radiat. Isotopes*, 25, 19, 1974.
79. **Loose-Wagenbach, I. and Clausnitzer, G.**, A system for production and transport of carbon-11 isotopes, *Nucl. Instr. Meth.*, 150, 345, 1978.
80. **Fares, Y., Demichele, D. W., Goeschl, J. D., and Baltuskonis, D. A.**, Continuously produced, high specific activity ^{11}C for studies of photosynthesis, transport and metabolism, *Int. J. Appl. Radiat. Isotopes*, 29, 431, 1978.
81. **Lamb, J. F., James, R. W., and Winchell, H. S.**, Recoil synthesis of high specific activity ^{11}C-cyanide, *Int. J. Appl. Radiat. Isotopes*, 22, 475, 1971.
82. **Finn, R. D., Christman, D. R., Ache, H. J., and Wolf, A. P.**, The preparation of cyanide-^{11}C for use in the synthesis of organic radiopharmaceuticals II, *Int. J. Appl. Radiat. Isotopes*, 22, 735, 1971.
83. **Christman, D. R., Finn, R. D., Karlstrom, K. I., and Wolf, A. P.**, Production of carrier-free $H^{11}CN$ for medical use and radiopharmaceutical syntheses, *J. Nucl. Med.*, 14, 864, 1973.
84. **Christman, D. R., Hoyte, R. M., and Wolf, A. P.**, Organic radiopharmaceuticals labelled with isotopes of short half-life I: ^{11}C-1-dopamine hydrochloride, *J. Nucl. Med.*, 11, 474, 1970.
85. **Cramer, R. D. and Kistiakowsky, G. B.**, The synthesis of radioactive lactic acid, *J. Biol. Chem.*, 137, 549, 1941.
86. **Loftfield, R. B.**, Preparation of ^{14}C-labeled hydrogen cyanide, alanine and glycine, *Nucleonics*, 1, 54, 1947.
87. **Hayes, R. L., Washburn, L. C., Wieland, B. W., Sun, T. T., Turtle, R. R., and Butler, T. A.**, Carboxyl-labelled ^{11}C-1-aminocyclopentane-carboxylic acid, a potential agent for cancer detection, *J. Nucl. Med.*, 17, 748, 1976.
88. **Nickles, R. J., Gatley, S. J., Madsen, M. T., Hichwa, R. D., Martin, J. L., Simpkin, D. J., and Halama, J. R.**, High yield synthesis of the sequence nitrogen-13 nitrogen gas, nitrogen-13 nitrous oxide, carbon-11 acetylene, oxygen-15 water, oxygen-14 water and fluorine-17 ion for regional cerebral blood flow studies, *J. Label. Comp. Radiopharm.*, 16, 90, 1979.
89. **Myers, W. G.**, ^{11}C-acetylene, *J. Nucl. Med.*, 13, 699, 1972.
90. **Vaalburg, W., Reiffers, S., Beerling, H. D., Pratt, J. J., Woldring, M. G., and Wynberg, H.**, The preparation of carbon-11 labelled 17α-ethynylestradiol, *J. Label. Comp. Radiopharm.*, 13, 200, 1977.
91. **Wolf, A. P., Christman, D. R., Fowler, J. S., and Lambrecht, R. M.**, Synthesis of radiopharmaceuticals and labelled compounds using short-lived isotopes, in *Radiopharmaceuticals and Labelled Compounds*, Vol. I, International Atomic Energy Agency, Vienna, 1973, 345.
92. **Machulla, H. J., Laufer, P., and Stocklin, G.**, Radioanalytical quality control of ^{11}C, ^{18}F and ^{123}I-labelled compounds and radiopharmaceuticals, *J. Radioanal. Chem.*, 32, 381, 1976.
93. **Crouzel, C., Sejourne, C., and Comar, D.**, Production of ^{11}C-acetylene by methane pyrolysis, *Int. J. Appl. Radiat. Isotopes*, 30, 566, 1979.
94. **Comar, D., Mazière, M., and Crouzel, C.**, Synthese et metabolisme de molecules radiopharmaceutiques marquées au carbon-11, in *Radiopharmaceuticals and Labelled Compounds*, Vol. I, International Atomic Energy Agency, Vienna, 1973, 461.
95. **Marazano, C., Mazière, M., Berger, G., and Comar, D.**, Synthesis of methyl iodide-^{11}C and formaldehyde-^{11}C, *Int. J. Appl. Radiat. Isotopes*, 28, 49, 1977.
96. **Långström, B. and Lundqvist, H.**, The preparation of ^{11}C-methyl iodide and its use in the synthesis of ^{11}C-methyl-L-methionine, *Int. J. Appl. Radiat. Isotopes*, 27, 357, 1976.
97. **Wagner, R. and Stöcklin, G.**, In target preparation of $^{11}CH_3I$ by ^{11}C-recoil reactions in the N_2/HI system, *J. Label. Comp. Radiopharm.*, 18, 189, 1981.
98. **Wagner, R.**, Zur Chemie von ^{11}C-Rückstossatomen in Alkylhalogeniden und Halogenwasserstoffen, Jül-1656, *Kernforschungsanlage*, 1980.
99. **Palmer, A. J.**, The preparation of ^{11}C-methyl labelled 1,1'-dimethyl-4,4'-dipyridinium diiode, *J. Label. Comp. Radiopharm.*, 14, 27, 1978.

100. **Iwata, R., Ido, T., Saji, H., Suzuki, K., Yoshikawa, K., Tamate, K., and Kasida, Y.**, A remote-controlled synthesis of ^{11}C-iodemethane for the practical preparation of ^{11}C-labelled radiopharmaceuticals, *Int. J. Appl. Radiat. Isotopes*, 30, 194, 1979.
101. **Långström, B. and Stridsberg, B.**, Syntheses of racemic (1-^{11}C)-alanine and partially resolved (3-^{11}C)-alanine, *Int. J. Appl. Radiat. Isotopes*, 30, 151, 1979.
102. **Reiffers, S., Vaalburg, W., Wiegman, T., Beerling, H. D., Paans, A. M. J., Woldring, M. G., and Wijnberg, H.**, ^{11}C-methyllithium — a new synthetic tool in radiopharmaceutical chemistry, *J. Label. Comp. Radiopharm.*, 16, 56, 1979.
103. **Reiffers, S., Vaalburg, W., Wiegman, T., Wijnberg, H., and Woldring, M. G.**, Carbon-11 labelled methyllithium as methyl donating agent: the addition to 17- keto steroids, *Int. J. Appl. Radiat. Isotopes*, 31, 535, 1980.
104. **Långström, B.**, On the Synthesis of ^{11}C-Compounds, Ph.D. thesis, University of Uppsala, Sweden, 1980.
105. **Straatmann, M. G. and Welch, M. J.**, A general method for labeling proteins with ^{11}C, *J. Nucl. Med.*, 16, 425, 1975.
106. **Marazano, C., Mazière, M., Berger, G., and Comar, D.**, Synthesis of methyl iodide-^{11}C and formaldehyde-^{11}C, *Int. J. Appl. Radiat. Isotopes*, 28, 49, 1977.
107. **Crawford, E. J., Christman, D. R., Atkins, H., Friedkin, M., and Wolf, A. P.**, Scintigraphy with positron-emitting compounds I; carbon-11 labeled thymidine and thymidylate, *Int. J. Nucl. Med. Biol.*, 5, 61, 1978.
108. **Calvin, M., Heidelberger, C., Reid, J. C., Tolbert, B. M., and Yankwich, P. F.**, *Isotopic Carbon*, John Wiley & Sons, New York, 1949, 155.
109. **Brinkman, G. A., Hass-Lisewska, I., Veenboer, J. Th., and Lindner, L.**, Preparation of ^{11}COCl$_2$, *Int. J. Appl. Radiat. Isotopes*, 29, 701, 1978.
110. **Christman, D. R., Finn, R. D., and Wolf, A. P.**, A novel synthesis of carbon-11 labeled phosgene, in 10th Hot Atom Chem. Symp., Loughborough, 1979.
111. **Roeda, D. and Westera, G.**, A novel method for the production of ^{11}C-phosgene, *J. Label. Comp. Radiopharm.*, 18, 11, 1981.
112. **Roeda, D., Crouzel, C., and Van Zanten, B.**, The production of ^{11}C-phosgene without added carrier, *Radiochem. Radioanal. Lett.*, 33, 175, 1978.
113. **Lederer, C. M. and Shirley, V. S.**, *Table of Isotopes*, 7th ed., John Wiley & Sons, New York, 1978, 16.
114. **Wieland, B. W., Highfill, R. R., and King, P. H.**, Proton accelerator targets for the production of ^{11}C, ^{13}N, ^{15}O and ^{18}F, *IEEE Trans. Nucl. Sci.*, NS-26, 1713, 1979.
115. **Vaalburg, W., Kamphuis, J. A. A., Beerling-van der Molen, H. D., Reiffers, S., Rijskamp, A., and Woldring, M. G.**, An improved method for the cyclotron production of ^{13}N-labelled ammonia, *Int. J. Appl. Radiat. Isotopes*, 26, 316, 1975.
116. **Austin, S. M., Galonsky, A., Bortins, J., and Wolk, C. P.**, A batch process for the production of ^{13}N-labelled nitrogen gas, *Nucl. Instr. Meth.*, 126, 373, 1975.
117. **McNaughton, G. S. and More, R. D.**, The use of a 3MV Van de Graaff accelerator for the production of ^{13}N-labelled ammonium and nitrate ions for biological experiments, *Int. J. Appl. Radiat. Isotopes*, 30, 489, 1979.
118. **Tilbury, R. S., Gelbard, A. S., McDonald, J. M., Benna, R. S., and Laughlin, J. S.**, A compact in-house multi-particle cyclotron for production of radionuclides and labeled compounds for medical use, *IEEE Trans. Nucl. Sci.*, NS-16, 1729, 1979.
119. **Valentin, L., Albony, G., Cohen, J. P., and Gusa how, M.**, Functions d'excitation des reactions (p,pn) et (p, 2p2n) dans les noyaux legers entre 15 et 155 MeV, *J. Phys. (Paris)*, 25, 704, 1964.
120. **Valentin, L.**, Reaction (p,n) et (p,pn) Induites a Moyenne Energie sur des Noyaux Legers, *Nucl. Phys.*, 62, 81, 1962.
121. **Albouy, G., Cohen, J. P., Gusakow, M., Poffe, N., Sergalle, H., and Valentin, L.**, Spallation de l'Oxygene par des Protons de 20 á 150 MeV, *Phys. Lett.*, 2, 306, 1962.
122. **Hill, H. A., Haase, E. L., and Knudsen, D. B.**, High-resolution measurements of the ^{16}O(p,α)^{13}N excitation function, *Phys. Rev.*, 123, 1301, 1961.
123. **Wilkinson, D. H.**, Reaction ^{12}C(d,n)^{13}N groundstate and ^{12}C(d,t)^{11}C up to Ed = 20 MeV, *Phys. Rev.*, 100, 32, 1955.
124. **Hahn, R. L. and Ricci, E.**, Interaction of ^{3}He particles with ^{9}Be, ^{12}C, ^{16}O and ^{19}F, *Phys. Rev.*, 146, 650, 1966.
125. **Lathrop, K. A., Harper, P. V., Rich, B., Dinwoodie, R., Krizek, H., Lembares, N., and Gloria, I.**, Rapid incorporation of short-lived cyclotron-produced radionuclides into radiopharmaceuticals, in *Radiopharmaceuticals and Labelled Compounds*, Vol. 1, International Atomic Energy Agency, Vienna, 1973, 471.

126. **Parks, N. J. and Krohn, K. A.**, The synthesis of ^{13}N-labeled ammonia, dinitrogen, nitrate and nitrite using a single cyclotron target system, *Int. J. Appl. Radiat. Isotopes*, 29, 754, 1978.
127. **Lindner, L., Helmer, J., and Brinkman, G. A.**, Water "loop" target for the in-cyclotron production of ^{13}N by the reaction $^{16}O(p,\alpha)^{13}N$, *Int. J. Appl. Radiat. Isotopes*, 30, 506, 1979.
128. **Skokut, T. A., Wolk, C. P., Thomas, J., Meeks, J. C., Shaffer, P. W., and Chien, W. S.**, Initial organic products of assimilation of nitrogen-13 ammonium and nitrogen-13 nitrate by tobacco cells cultured on different sources of nitrogen, *Plant Physiol.*, 62, 299, 1978.
129. **Welch, M. J. and Straatmann, M. G.**, The reaction of recoil ^{13}N atoms with some organic compounds in the solid and liquid phase, *Radiochim. Acta*, 20, 124, 1973.
130. **Gersberg, R., Krohn, K., Peek, N., and Goldman, C. R.**, Denitrification studies with ^{13}N-labeled nitrate, *Science*, 192, 1229, 1976.
131. **Tilbury, R. S. and Dahl, J. R.**, ^{13}N-species formed by proton irradiation of water, *Radiat. Res.*, 79, 22, 1979.
132. **Krizek, H., Lembares, N., Dinwoodie, R., Gloria, I., Lathrop, K. A., and Harper, P. V.**, Production of radiochemically pure $^{13}NH_3$ for biomedical studies using the $^{16}O(p,\alpha)^{13}N$ reaction, *J. Nucl. Med.*, 14, 629, 1973.
133. **Gelbard, A. S., Clarke, L. P., McDonald, J. M., Monahan, W. G., Tilbury, R. S., Kuo, T. Y. T., and Laughlin, J. S.**, Enzymatic synthesis and organ distribution studies with ^{13}N-labelled L-Glutamine and L-Glutamic acid, *Radiology*, 116, 127, 1975.
134. **Majumdar, C., Lathrop, K. A., and Harper, P. V.**, Synthesis and analysis of ^{13}N-asparagine for myocardial scanning, *J. Label. Comp. Radiopharm.*, 13, 206, 1977.
135. **Ido, T. and Iwata, R.**, Fully automated synthesis of $^{13}NH_3$, *J. Label. Comp. Radiopharm.*, 18, 244, 1981.
136. **Welch, M. J. and Lifton, J. F.**, The fate of nitrogen-13 formed by the $^{12}C(d,n)^{13}N$ reaction in inorganic carbides, *J. Am. Chem. Soc.*, 93, 3385, 1971.
137. **Carter, C. C., Lifton, J. F., and Welch, M. J.**, Organ uptake and blood pH and concentration effects of ammonia in dogs determined with ammonia labeled with 10 minute half-lived nitrogen-13, *Neurology*, 23, 204, 1973.
138. **Tilbury, R. S., Dahl, J. R., Monahan, W. G., and Laughlin, J. S.**, The production of ^{13}N-labeled ammonia for medical use, *Radiochem. Radioanal. Lett.*, 8, 317, 1971.
139. **Gelbard, A. S., Hara, T., Tilbury, R. S., and Laughlin, J. S.**, Recent aspects of cyclotron production of medically useful isotopes, in *Radiopharmaceuticals and Labelled Compounds*, Vol. I, International Atomic Energy Agency, Vienna, 1973, 161.
140. **Ruben, S., Hassid, W. Z., and Kamen, M. D.**, Radioactive nitrogen in the study of N_2 fixation by non-leguminous plants, *Science*, 91, 578, 1940.
141. **Buckingham, P. D. and Forse, G. R.**, The preparation and processing of radioactive gases for clinical use, *Int. J. Appl. Radiat. Isotopes*, 14, 439, 1963.
142. **Buckingham, P. D. and Clark, J. C.**, Nitrogen-13 solutions for research studies in pulmonary physiology, *Int. J. Appl. Radiat. Isotopes*, 23, 5, 1972.
143. **Nicholas, D. J. D., Silvester, D. J., and Fowler, J. F.**, Use of radioactive nitrogen in studying nitrogen fixation in bacterial cells and their extracts, *Nature (London)*, 189, 634, 1961.
144. **Clark, J. C. and Buckingham, P. D.**, *Short-Lived Radioactive Gases for Clinical Use*, Butterworths, London, 1975, chap. 6.
145. **Del Fiore, G., Depresseux, J. C., Bartsch, P., Quaglia, L., and Peters, J. M.**, Production of oxygen-15, nitrogen-13 and carbon-11 and their low molecular weight derivatives for biomedical application, *Int. J. Appl. Radiat. Isotopes*, 30, 543, 1979.
146. **Ritchie, A. I. M.**, The production of the radioisotopes ^{11}C, ^{13}N and ^{15}O using the deuteron beam from a 3 MeV van de Graaff accelerator, *Nucl. Instr. Meth.*, 64, 181, 1968.
147. **Welch, M. J.**, Production of active molecular nitrogen by the reaction of recoil nitrogen-13, *Chem. Comm.*, 21, 1354, 1968.
148. **Brownell, G. L.**, Quantitative dynamic studies using short-lived radioisotopes and positron detection, in *Dynamic Studies with Radioisotopes in Medicine*, IAEA, Vienna, 1971, 161.
149. **Greene, R., Hoop, B., and Kazemi, H.**, Use of ^{13}N in studies of airway closure and regional ventilation, *J. Nucl. Med.*, 12, 719, 1971.
150. **Crouzel, C. and Comar, D.**, Production of nitrogen-13 solutions for injection by means of a medical cyclotron, *Radiochem. Radioanal. Lett.*, 20, 273, 1975.
151. **Jones, S. C., Bucelewicz, W. M., Brisette, R. A., Subramanyam, R., and Hoop, B.**, Production of ^{13}N-molecular nitrogen for pulmonary positron scintigraphy, *Int. J. Appl. Radiat. Isotopes*, 28, 25, 1977.
152. **Austin, S. M., Galonski, A., Bortins, J., and Wolk, C. P.**, A batch process for the production of ^{13}N-labeled nitrogen gas, *Nucl. Instr. Meth.*, 126, 373, 1975.
153. **Suzuki, K. and Iwata, R.**, A novel method for the production of ^{13}NN proton irradiation of an aqueous solution of ammonia, *Radiochem. Radioanal. Lett.*, 28, 263, 1977.

154. **Vaalburg, W., Reiffers, S., Paans, A. M. J., Beerling, H. D., Nickles, R. J., Krizek, H., and Harper, P. V.**, N-13 ammonia oxidation to molecular nitrogen for pulmonary function studies, *J. Nucl. Med.*, 18, 638, 1977.
155. **Vaalburg, W., Steenhoek, A., Paans, A. M. J., Peset, R., Reiffers, S., and Woldring, M. G.**, Production of ^{13}N-labelled molecular nitrogen for pulmonary function studies, *J. Label. Comp. Radiopharm.*, 18, 303, 1981.
156. **Nickles, R. J., Gatley, S. J., Hichwa, R. D., Simpkin, D. J., and Martin, J. L.**, The synthesis of ^{13}N-labelled nitrous oxide, *Int. J. Appl. Radiat. Isotopes*, 29, 225, 1978.
157. **Krizek, H.**, Private communication.
158. **Ogg, R. A.**, Isotopic nitrogen exchange between nitrogen pentoxide and dioxide, *J. Chem. Phys.*, 15, 613, 1947.
159. **Lederer, C. M. and Shirley, V. S.**, *Table of Isotopes*, 7th ed., John Wiley & Sons, New York, 1978, 19.
160. **Lederer, C. M. and Shirley, V. S.**, *Table of Isotopes*, 7th ed., John Wiley & Sons, New York, 1978, 17.
161. **Beaver, J. E., Finn, R. D., and Hupf, H. B.**, A new method for the production of high concentration oxygen-15 labeled carbon dioxide with protons, *Int. J. Appl. Radiat. Isotopes*, 27, 195, 1976.
162. **Clark, J. C. and Buckingham, P. D.**, *Short-Lived Radioactive Gases for Clinical Use*, Butterworth, London, 1975, 122.
163. **Dahl, J. R., Tilbury, R. S., Russ, G. A., and Bigler, R. E.**, The preparation of oxygen-14 for biomedical use, *Radiochem. Radioanal. Lett.*, 26, 107, 1976.
164. **Nickles, R. J., Kiuru, A. J., Schuster, S. M., and Holden, J. E.**, A catalytic generator for the production of $H_2^{14}O$ and $H_2^{15}O$, *Prog. Nucl. Med.*, 4, 72, 1978.
165. **Vera Ruiz, H. and Wolf, A. P.**, Excitation function for ^{15}O production via the $^{14}N(d,n)^{15}O$ reaction, *Radiochim. Acta*, 24, 65, 1977.
166. **Albouy, G., Cohen, J. P., Gusakow, M., Poffe, N., Sergolle, M., and Valentin, L.**, Spallation de l'Oxygene par des Protons de 20 a 150 MeV, *Phys. Lett.*, 2, 306, 1962.
167. **Valentin, L.**, Reaction (p,n) et (p,pn) Induites a Moyenne energie sur des Noyaux Legers, *Nucl. Phys.*, 62, 81, 1965.
168. **Newson, H. W.**, Transmutation functions at high bombarding energies, *Phys. Rev.*, 51, 620, 1937.
169. **Retz Schmidt, T. and Weil, J. L.**, Excitation curves and angular distributions for $^{14}N(d,n)^{15}O$, *Phys. Rev.*, 119, 1079, 1960.
170. **Wohlleben, K. and Schuster, E.**, Der totale Wirkungsquerschnitt der Reaktionen $^{10}B(d,n)^{11}C$, $^{14}N(d,n)^{15}O$ und $^{16}O(d,n)^{17}F$ bis 3,2 MeV, *Radiochim. Acta*, 12, 75, 1969.
171. **Buckingham, P. D. and Forse, G. R.**, The preparation and processing of radioactive gases for clinical use, *Int. J. Appl. Radiat. Isotopes*, 14, 439, 1963.
172. **Welch, M. J. and Ter-Pogossian, M. M.**, Preparation of short half-lived radioactive gases for medical studies, *Radiat. Res.*, 36, 580, 1968.
173. **Clark, J. C. and Buckingham, P. D.**, *Short-Lived Radioactive Gases for Clinical Use*, Butterworths, London, 1975, chap. 5.
174. **Subramanyam, R., Bucelewicz, W. M., Hoop, B., and Jones, S. C.**, A system for oxygen-15 labeled blood for medical applications, *Int. J. Appl. Radiat. Isotopes*, 28, 21, 1977.
175. **Beaver, J. E., Finn, R. D., and Hupf, H. B.**, A new method for the production of high concentration oxygen-15 labeled carbon dioxide with protons, *Int. J. Appl. Radiat. Isotopes*, 27, 195, 1976.
176. **Ruiz, H. V. and Wolf, A. P.**, Direct synthesis of oxygen-15 labelled water of high specific activity, *J. Label. Comp. Radiopharm.*, 15, 185, 1978.
177. **Harper, P. V. and Wickland, T.**, Oxygen-15 labelled water for continuous intravenous administration, *J. Label. Comp. Radiopharm.*, 18, 186, 1981.
178. **Welch, M. J., Lefton, J. F., and Ter-Pogossian, M. M.**, Preparation of millicurie quantities of oxygen-15 labeled water, *J. Label. Comp.*, 5, 168, 1969.
179. **Russ, G. A., Bigler, R. E., Dahl, J. R., Kostick, J., McDonald, J. M., Tilbury, R. S., and Laughlin, J. S.**, Regional imaging with oxygen-14, *IEEE Trans. Nucl. Sci.*, NS-23, 573, 1976.
180. **Kamen, M. D.**, Short-lived radioactive carbon-11, in *Radioactive Tracers in Biology*, Academic Press, New York, 1959, chap. 7.
181. **Iwata, R., Ido, T., and Tominaga, T.**, The production of ^{11}C-guanidine by the proton irradiation of the liquid ammonia-nitrous oxide system and its use in the synthesis of ^{11}C-pyrimidines, *Int. J. Appl. Radiat. Isotopes*, 32, 303, 1981.
182. **Rössler, K., Vogt, M., and Stöcklin, G.**, Production of ^{11}C-precursors via the $^{14}N(p,\alpha)^{11}C$ reaction in solid ammonium halides at higher doses, *J. Label. Comp. Radiopharm.*, 18, 190, 1981.
183. **Långström, B. and Sjöberg, S.**, The synthesis of aliphatic and aromatic hydrocarbons containing methyl groups labelled with ^{11}C, *J. Label. Comp. Radiopharm.*, 18, 671, 1981.

184. **Machulla, H. J. and Dutschka, K.,** Carbon-11 labelled radiopharmaceuticals: synthesis and high pressure liquid chromatography of carbon-11 nicotine-acid amide, *J. Label. Comp. Radiopharm.*, 16, 287, 1979.
185. **Pike, V. W., Eakins, M. N., Allen, R. M., and Selwin, A. P.,** Preparation of ^{11}C-labelled acetate for the study of myocardial metabolism by emission-computerized axial tomography, *J. Label. Comp. Radiopharm.*, 18, 249, 1981.
186. **Fowler, J. S., Gallagher, B. M., MacGregor, R. R., and Wolf, A. P.,** Carbon-11 labeled aliphatic amines in lung uptake and metabolism studies: potential for dynamic measurements in vivo, *J. Pharm. Exp. Ther.*, 198, 133, 1976.
187. **Weiss, E. S., Hoffman, E. J., Phelps, E. M., Welch, M. J., Henry, P. D., Ter-Pogossian, M. M., and Sobel, E. R.,** External detection and visualization of myocardial ischemia with ^{11}C-substrates in vitro and in vivo, *Circ. Res.*, 39, 24, 1976.
188. **Machulla, H. J., Stöcklin, G., Kupfernagel, Ch., Freundlieb, Ch., Höck, A., Vyska, K., and Feinendegen, L. E.,** Comparative evaluation of fatty acids labelled with C-11, Cl-34m, Br-77, and I-123 for metabolic studies of the myocardium: concise communication, *J. Nucl. Med.*, 19, 298, 1978.
189. **Raichle, M. E., Eichling, J. O., Straatmann, M. G., Welch, M. J., Larson, K. B., and Ter-Pogossian, M. M.,** Blood-brain barrier permeability of ^{11}C-labeled alcohols and ^{15}O-labeled water, *Am. J. Physiol.*, 230, 543, 1976.
190. **Keinänen, M., Kuikka, J., Heselius, S. J., Suolinna, E. M., Solin, O., Långström, B., and Näntö, V.,** Transcapilary exchange and distribution of carbon-11 labelled ethanol in the isolated perfused rat liver, *Acta Physiol. Scand.*, 110, 39, 1980.
191. **Dischino, D. D., Wittmer, S. L., Raichle, M. E., and Welch, M. J.,** Synthesis and cerebral extraction studies of carbon-11 labelled ethers, *J. Label. Comp. Radiopharm.*, 18, 238, 1981.
192. **Kutzman, R. S., Meyer, G. J., and Wolf, A. P.,** The distribution and excretion of inhaled carbon-11 benzaldehyde by the rat after acute exposure, *Pharmacologist*, 20, 145, 1978.
193. **Tang, D. Y., Lipman, A., Meyer, G. J., Wan, C. N., and Wolf, A. P.,** Carbon-11 labelled octanal and benzaldehyde, *J. Label. Comp. Radiopharm.*, 16, 435, 1979.
194. **Reiffers, S., Beerling, H. D., Vaalburg, W., Ten Hoeve, W., Woldring, M. G., and Wynberg, H.,** Synthesis of carbon-11 labeled DOPA, *J. Label. Comp. Radiopharm.*, 13, 198, 1977.
195. **Reiffers, S., Van der Molen, H. D., Vaalburg, W., Ten Hoeve, W., Paans, A. M. J., Korf, J., Woldring, M. G., and Wynberg, H.,** Rapid synthesis and purification of carbon-11 labeled DOPA: a potential agent for brain studies, *Int. J. Appl. Radiat. Isotopes*, 28, 955, 1977.
196. **Långström, B., Stridsberg, B.,** Synthesis of racemic 1-^{11}C-alanine and partially resolved 3-^{11}C-alanine, *Int. J. Appl. Radiat. Isotopes*, 30, 151, 1979.
197. **Washburn, L. C., Sunn, T. T., Rafter, J. J., and Hayes, R. L.,** Carbon-11 labeled amino-acids for pancreas visualization, *J. Nucl. Med.*, 17, 557, 1976.
198. **Washburn, L. C., Wieland, B. W., Sun, T. T., Hayes, R. L., and Butler, T. A.,** 1-^{11}C-Dl-valine, a potential pancreas-imaging agent, *J. Nucl. Med.*, 19, 77, 1978.
199. **Hayes, R. L., Washburn, L. C., Wieland, B. W., Sun, T. T., Anon, J. B., Butler, T. A., and Callahan, A. P.,** Synthesis and purification of ^{11}C-carboxyl-labeled amino acids, *Int. J. Appl. Radiat. Isotopes*, 29, 186, 1978.
200. **Washburn, L. C., Sun, T. T., Wieland, B. W., and Hayes, R. L.,** Carbon-11 racemic tryptophan, a potential pancreas imaging agent for positron tomography, *J. Nucl. Med.*, 18, 638, 1977.
201. **Washburn, L. C., Sun, T. T., Byrd, B. L., Hayes, R. L., and Butler, T. A.,** DL-carboxyl-^{11}C-tryptophan, a potential agent for pancreatic imaging: production and preclinical investigations, *J. Nucl. Med.*, 20, 857, 1979.
202. **Wieland, B. W., Washburn, L. C., Turtle, R. R., Hayes, R. L., and Butler, T. A.,** Development of cyclotron targetry and remote radiochemical techniques for the continuous large-scale production of carbon-11 labeled amino acids, *J. Label. Comp. Radiopharm.*, 13, 202, 1977.
203. **Washburn, L. C., Sun, T. T., Byrd, B. L., Hayes, R. L., and Butler, T. A.,** 1-aminocyclobutane-^{11}C-carboxylic acid, a potential tumor-seeking agent, *J. Nucl. Med.*, 20, 1055, 1979.
204. **Hübner, K. F., Andrews, G. A., Washburn, L. C., Wieland, B. W., Gibbs, W. D., Hayes, R. L., Butler, T. A., and Winebrenner, J. D.,** Tumor localization with 1-aminocyclopentane-^{11}C-carboxylic acid: preliminary clinical trials with single photon detection, *J. Nucl. Med.*, 18, 1215, 1977.
205. **Comar, D., Cartron, J. C., Mazière, M., and Marazano, C.,** Labelling and metabolism of methionine-methyl-^{11}C, *Eur. J. Nucl. Med.*, 1, 11, 1976.
206. **Berger, G., Mazière, M., Knipper, R., Prenant, C., and Comar, D.,** Automated synthesis of carbon-11 labeled radiopharmaceuticals imipramine, chlorpromazine, nicotine and methionine, *Int. J. Appl. Radiat. Isotopes*, 30, 393, 1979.
207. **Långström, B., Sjöberg, S., and Ragnarsson, U.,** A rapid and convenient method for specific ^{11}C-labelling of synthetic polypeptides containing methionine, *J. Label. Comp. Radiopharm.*, 18, 479, 1981.
208. **Kloster, G. and Laufer, P.,** Enzymatic synthesis and chromatographic purification of L-3-^{11}C-lactic acid via DL-3-^{11}C-alanine, *J. Label. Comp. Radiopharm.*, 17, 889, 1980.

209. **Cohen, M. B., Spolter, L., Chang, C. C., and MacDonald, N. S.,** Enzymatic synthesis of carbon-11 pyruvic acid, lactic acid and L-alanine, *J. Nucl. Med.,* 19, 701, 1978; *J. Label. Comp. Radiopharm.,* 16, 63, 1979.
210. **Laughlin, J. S., Benua, R. S., Gelbard, A. S., Reiman, R., Bigler, R. E., Schmall, B., Rosen, G., Hopfan, S., Allen, J., Helson, L., Dahl, J. R., and Lee, R.,** Report on compounds labelled with N-13 or C-11 used in cancer metabolic studies with quantitative two-dimensional scanning and Pet tomography, in *Medical Radionuclide Imaging,* IAEA, Vienna, 1981.
211. **Cohen, M. B., Spolter, L., Chang, C. C., and MacDonald, N. S.,** Six-step enzymatic synthesis of C-11 L-glutamic acid, *Clin. Nucl. Med.,* 5, S15, 1980.
212. **Winstead, M. B., Parr, S. J., Rogal, M. J., Brookman, P. S., Lubcher, R., Khentigan, A., Lin, T. H., Lamb, J. F., and Winchell, H. S.,** Relationship of molecular structure to in vivo scintigraphic distribution patterns of carbon 11 labeled compounds: 3, ^{11}C-Hydantoins, *J. Med. Chem.,* 19, 279, 1976.
213. **Stavchansky, S. A., Tilbury, R. S., McDonald, J. M., Ting, C. T., and Kostenbauder, H. B.,** In vivo distribution of carbon-11 phenytoin and its major metabolite, and their use in scintigraphic imaging, *J. Nucl. Med.,* 19, 936, 1978.
214. **Roeda, D., Crouzel, C., Van der Jagt, P. J., Van Zanten, B., and Comar, D.,** Synthesis of ^{11}C-urea for medical use, *Int. J. Appl. Radiat. Isotopes,* 31, 549, 1980.
215. **Kloster, G.,** Müller-Platz, Synthesis, chromatography and tissue distribution of 1-^{11}C-hexobarbital, *J. Label. Comp. Radiopharm.,* 18, 242, 1981.
216. **Winstead, M. B., Chuen-Ing Chern, Tz-Hong Lin, Khentigan, A., Lamb, J. F., and Winchell, H. S.,** Synthesis and preliminary scintigraphic evaluation of in vivo distribution of ^{11}C-hydroxyurea isohydroxyurea and ^{11}C-cyanate, *Int. J. Appl. Radiat. Isotopes,* 29, 443, 1978.
217. **Fowler, J. S., Gallagher, B. M., MacGregor, R. R., Wolf, A. P., Ansari, A. N., Atkins, H. L., and Slatkin, D. N.,** Radiopharmaceuticals XIX, ^{11}C-labeled octylamine, a potential diagnostic agent for lung structure and function, *J. Nucl. Med.,* 17, 752, 1976.
218. **Fowler, J. S., MacGregor, R. R., and Wolf, A. P.,** Radiopharmaceuticals, 16, Halogenated dopamine analogs. Synthesis and radiolabeling of 6-iododopamine and tissue distribution studies in animals, *J. Med. Chem.,* 19, 356, 1976.
219. **Winstead, M. B., Chuen-Ing Chern, Tz-Hong Lin, Khentigan, A., Lamb, J. F., and Winchell, H. S.,** Synthesis and preliminary scintigraphic evaluation of in vivo distribution of ^{11}C-lactic acid and ^{11}C-lactonitrile, *Int. J. Appl. Radiat. Isotopes,* 29, 69, 1978.
220. **Winstead, M. B. and Dougherty, D. A.,** Relationship of molecular structure to in vivo scintigraphic distribution of carbon-11 labelled compounds, 4, Carbon-11 labeled mandelonitriles, mandelic acids, and their esters, *J. Med. Chem.,* 21, 215, 1978.
221. **Jay, M., Digenis, G. A., Chaney, J. E., Washburn, L. C., Byrd, B. L., Hayes, R. L., and Callahan, A. P.,** Synthesis and brain uptake of carbon-11 phenethylamine, *J. Label. Comp. Radiopharm.,* 18, 237, 1981.
222. **Winstead, M. B. and Podlesny, E. J.,** Preparation, scintigraphic evaluation and dissociation study of carbon-11 2-N-phenylethyl-aminoalkane-nitrile hydrochlorides, *J. Label. Comp. Radiopharm.,* 13, 215, 1977.
223. **Winstead, M. B., Podlesny, E. J., and Winchell, H. S.,** Preparation and scintigraphic evaluation of carbon-11 labeled 2-N-phenethylaminoalkane-nitrils, *Nucl. Med.,* 17, 277, 1978.
224. **Winstead, M. B., Dischino, D. D., and Winchell, H. S.,** Concentration of activity in brain following administration of carbon-11 labeled alpha-p-iodo-anilinophenyl acetonitrile, *Int. J. Appl. Radiat. Isotopes,* 30, 293, 1979.
225. **Winstead, M. B., Dischino, D. D., and Winchell, H. S.,** Relationship of molecular structure to the in vivo distribution of carbon-11 labelled compounds V: carbon-11 labeled N-alkyl-p-iodobenzenesulfonamides, *Nucl. Med.,* 18, 142, 1979.
226. **Winstead, M. B., Dischino, D. D., Munder, N. A., Walsh, C., and Winchell, H. S.,** Relationship of molecular structure to in vivo distribution of carbon-11 labeled compounds. VI. Carbon-11 labeled aliphatic diamines, *Eur. J. Nucl. Med.,* 5, 165, 1980.
227. **Mestelan, G., Crouzel, C., and Comar, D.,** Synthesis and distribution kinetics in animals of α-^{11}C-3,4-dimethyoxyphenethyl-amine, *Int. J. Nucl. Med. Biol.,* 4, 185, 1977.
228. **Fowler, J. S., Gallagher, B. M., MacGregor, R. R., and Wolf, A. P.,** ^{11}C-serotonin: a tracer for the pulmonary endothelial extraction of serotonin, *J. Label. Comp. Radiopharm.,* 13, 194, 1977.
229. **Welch, M. J., Coleman, R. E., Straatmann, M. G., Asberry, B. E., Primeau, J. L., Fair, W. R., and Ter-Pogossian, M. M.,** Carbon-11 labeled methylated polyamine analogs: uptake in prostate and tumor in animals, *J. Nucl. Med.,* 18, 74, 1977.
230. **Mazière, M., Comar, D., Marazano, C., and Berger, G.,** Nicotine-^{11}C: synthesis and distribution kinetics in animals, *Eur. J. Nucl. Med.,* 1, 255, 1976.

231. **Berger, G., Mazière, M., Knipper, R., Prenant, C., and Comar, D.,** Automated synthesis of ^{11}C-labelled radiopharmaceuticals imipramine, chlorpromazine, nicotine and methionine, *Int. J. Appl. Radiat. Isotopes*, 30, 393, 1979.
232. **Mazière, M., Berger, G., and Comar, D.,** ^{11}C-labelled compounds in dynamic imaging studies of the brain, in *Medical Radionuclide Imaging*, Vol. 2, IAEA, Vienna, 1976.
233. **Mazière, M., Berger, G., and Comar, D.,** Optimization of ^{11}C-labeling of radiopharmaceuticals: ^{11}C-clomipramine as an example, *J. Label. Comp. Radiopharm.*, 13, 196, 1977.
234. **Mazière M., Berger, G., and Comar, D.,** ^{11}C-clomipramine: synthesis and analysis, *J. Radioanal. Chem.*, 45, 453, 1978.
235. **Mazière, M., Berger, G., Godot, J. M., Prenant, C., and Comar, D.,** Etorphine ^{11}C: a new tool for in vivo study of brain opiate receptors, *J. Label Comp. Radiopharm.*, 18, 15, 1981.
236. **Berger, G., Mazière, M., Marazano, C., and Comar, D.,** Carbon-11 labeling of the psychoactive drug O-methyl-bufotenine and its distribution in the animal organism, *Eur. J. Nucl. Med.*, 3, 101, 1978.
237. **Mazière, M., Godot, J. M., Berger, G., Prenant, C., and Comar, D.,** High specific activity carbon-11 labelling of benzodiazepines: diazepam and flunitrazepam, *J. Radioanal. Chem.*, 56, 229, 1980.
238. **Kloster, G., Röder, E., and Machulla, H. J.,** Synthesis, chromatography and tissue distribution of methyl-^{11}C-morphine and methyl-^{11}C-heroin, *J. Label. Comp. Radiopharm.*, 16, 441, 1979.
239. **Allen, D. R. and Beamier, P.,** Carbon-11 N-methylmorphine, *J. Label. Comp. Radiopharm.*, 16, 61, 1979.
240. **Ido, I. and Iwata, R.,** Caffeine-^{11}C, ephadrine-^{11}C and methylephedrine-^{11}C: synthesis and distribution in mice, *Radioisotopes*, 27, 451, 1978.
241. **Anwar, M., Mock, B., Lathrop, K. A., and Harper, P. V.,** The localization of carbon-11 labeled hexamethonium in cartilage, *J. Nucl. Med.*, 18, 638, 1977.
242. **Crouzel, C., Mestelan, G., Kraus, E., Lecomte, J. M., and Comar, D.,** Synthesis of a ^{11}C-labelled neuroleptic drug: pimozide, *Int. J. Appl. Radiat. Isotopes*, 31, 545, 1980.
243. **Vaalburg, W., Feenstra, A., Wiegman, T., Beerling, H. D., Reiffers, S., Talma, A., Woldring, M. G., and Wijnberg, H.,** Carbon-11 labelled moxestrol and 17α-methylestradiol as receptor binding radiopharmaceuticals, *J. Label. Comp. Radiopharm.*, 18, 100, 1981.
244. **Reiffers, S., Vaalburg, W., Wiegman, T., Wynberg, H., and Woldring, M. G.,** Carbon-11 labelled methyllithium as methyl donating agent: the addition of 17-keto steroids, *Int. J. Appl. Radiat. Isotopes*, 31, 535, 1980.
245. **Berger, G., Mazière, M., Prenant, C., and Comar, D.,** Synthesis of carbon-11 labelled acetone, *Int. J. Appl. Radiat. Isotopes*, 31, 577, 1980.
246. **Ehrin, E., Westman, E., Nilsson, S. O., Nillson, J. L. G., Widen, L., Greitz, T., Larsson, C. M., Tillberg, J. E., and Malmborg, P.,** A convenient method for production of ^{11}C-labeled glucose, *J. Label. Comp. Radiopharm.*, 17, 453, 1980.
247. **Feron, A., Peters, J. M., Del Fiore, G., Quaglia, L., Depresseux, J. C., and Bartsch, P.,** Photosynthesis of ^{11}C-labelled glucose with algae, *Eur. J. Nucl. Med.*, 5, A17, 1980.
248. **Shine, C. Y., MacGregor, R. R., Lade, R. E., Wan, C. N., and Wolf, A. P.,** The synthesis of 1-^{11}C-2-deoxy-D-glucose for measuring regional brain glucose metabolism in vivo, *J. Nucl. Med.*, 19, 676, 1978.
249. **Fowler, J. S., Lade, R. E., MacGregor, R. R., Shine, C., Wan, C. N., and Wolf, A. P.,** Agents for the armamentarium of regional metabolic measurement in vivo via metabolic trapping: ^{11}C-2-deoxy-D-glucose and halogenated derivatives, *J. Label. Comp. Radiopharm.*, 16, 7, 1979.
250. **Vyska, K., Höck, A., Freundlieb, C., Feinendegen, L. E., Kloster G., and Stöcklin, G.,** 3-^{11}C-methyl-glucose, a promising agent for in vivo assessment of function of myocardial cell membrane, *J. Nucl. Med.*, 21, P56, 1980.
251. **Kloster, G., Müller Platz, C., and Lauffer, P.,** 3-^{11}C-methyl-D-glucose, a potential agent for regional cerebral glucose utilization studies, in Proc. 17th. Int. Ann. Meeting. Soc. Nucl. Med., Innsbrück, 1979, 210.
252. **Crawford, E. J., Christman, D. R., Atkins, H., Friedkin, M., and Wolf, A. P.,** Scintigraphy with positron-emitting compounds-I: carbon-11 labeled thymidine and thymidylate, *Int. J. Nucl. Med. Biol.*, 5, 61, 1978.
253. **Spolter, L., Chang, C. C., Bobinet, D., Cohen, M. B., MacDonald, N. S., Flesher, A., and Takahashi, J.,** Enzymatic synthesis of carbon-11 citric acid, *J. Nucl. Med.*, 17, 558, 1976.
254. **Cohen, M. B., Spolter, L., Chang, C. C., Cook, J. S., and MacDonald, N. S.,** Enzymatic synthesis of ^{11}C-pyruvic acid and ^{11}C-L-lactic acid, *Int. J. Appl. Radiat. Isotopes*, 31, 45, 1980.
255. **Gatley, S. J., Crawford, J. S., Halama, J. S., Hichwa, R. D., Madsen, M. T., Martin, J. L., Nickles, R. J., and Simpkin, D. J.,** Synthesis of carbon-11 hippuric-acid from carbon-11 benzoic acid and glycine using rat liver mitochondria, *J. Label. Comp. Radiopharm.*, 16, 182, 1979.
256. **Pettit, W. A., Tilbury, R. S., Digenis, G. A., and Mortara, R. H.,** A convenient synthesis of ^{13}N-BCNU, *J. Label. Comp. Radiopharm.*, 13, 119, 1977.

257. **Pettit, W. A.**, Chemical considerations in the accelerator production of N-13: N-13-nitrosourea preparation, *IEEE Trans. Nucl. Sci.*, NS-26, 1718, 1979.
258. **Digenis, G. A., McQuinn, R. L., Freed, B., Tilbury, R. S., Reiman, R. E., and Cheng, Y. C.**, Preparation of ^{13}N-labelled streptozotocin and nitrosocarbaryl, *J. Label. Comp. Radiopharm.*, 16, 95, 1979.
259. **Krizek, H., Harper, P. V., and Mock, B.**, Adapting the old to new needs: nitrogen-13 labeled urea, *J. Label. Comp. Radiopharm.*, 13, 207, 1977.
260. **Gelbard, A. S., Benua, R. S., Reiman, R. E., McDonald, J. M., Vomero, J. J., and Laughlin, J. S.**, Imaging of the human heart after administration of L-(N-13)glutamate, *J. Nucl. Med.*, 21, 988, 1980.
261. **Baumgartner, F. J., Barrio, J. R., Henze, E., Schelbert, H., MacDonald, N. S., and Kuhl, D. E.**, Immobilized enzymes for the production of N-13 labeled L-amino acids, *Clin. Nucl. Med.*, 5, S16, 1980.
262. **Spolter, L., Cohen, M. B., Chang, C. C., and MacDonald, N. S.**, Semi-automated synthesis and purification of radiopharmaceutically pure ^{13}N-L-alanine in a continuous flow system, *J. Label. Comp. Radiopharm.*, 13, 204, 1977.
263. **Gelbard, A. S. and Cooper, A. J. L.**, Synthesis of ^{13}N-labeled aromatic L-amino acids by enzymatic transamination of ^{13}N-L-glutamic acid, *J. Label. Comp. Radiopharm.*, 16, 94, 1979.
264. **Elmaleh, D. R., Hnatowich, D. J., and Kulprathipanja, S.**, A novel synthesis of nitrogen-13 asparagine, *J. Label. Comp. Radiopharm.*, 16, 92, 1979.
265. **Reene, J. R.**, Nuclear data sheets for A = 52, *Nucl. Data Sheets*, 25, 235, 1978.
267. **Greene, M. W., Lebowitz, E., Richards, P., and Hillman, H.**, Production of ^{52}Fe for medical use, *Int. J. Appl. Radiat. Isotopes*, 21, 719, 1970.
268. **Saha, G. B. and Farrer, P. A.**, Production of ^{52}Fe by the ^{55}Mn(p,4n)^{52}Fe reaction for medical use, *Int. J. Appl. Radiat. Isotopes*, 22, 495, 1971.
269. **Dahl, J. R. and Tilbury, R. S.**, The use of a compact multi-particle cyclotron for the production of ^{52}Fe, ^{67}Ga, ^{111}In and ^{123}I for medical purpose, *Int. J. Appl. Radiat. Isotopes*, 23, 431, 1971.
270. **Atcher, R. W., Friedman, A. M., and Huizenga, J. R.**, Production of ^{52}Fe for use in a radionuclide generator system, *Int. J. Nucl. Med. Biol.*, 7, 75, 1980.
271. **Atcher, R. W., Friedman, A. M., Huizenga, J. R., Rayada, G. V. S., Silverstein, E. A., and Turner, D. A.**, Manganese-52m, A new short-lived, generator produced radionuclide: a potential tracer for positron tomography, *J. Nucl. Med.*, 21, 565, 1980.
272. **Chauncey, D. M., Schelbert, H. R., and Halpern, S. E.**, Tissue distribution studies with radioactive manganese: a potential agent for myocardial imaging, *J. Nucl. Med.*, 18, 933, 1977.
273. **Borg, D. C. and Cotzias, G. C.**, Manganese metabolism in man: rapid exchange of ^{56}Mn with tissue as demonstrated by blood clearance and liver uptake, *J. Clin. Invert.*, 27, 1269, 1958.
274. **Atkins, H. L., Som, P., Fairchild, R. G., Hui, J., Schachner, E., Goldman, A., and Ku, T.**, Myocardial positron tomography with manganese-52m, *Radiology*, 133, 769, 1979.
275. **Halbert, M. L.**, Nuclear data sheets for A = 62, *Nucl. Data Sheets*, 26, 5, 1979.
276. **Robinson, G. D. and Lee, A. W.**, Copper-62 radiopharmaceuticals, *J. Label. Comp. Radiopharm.*, 13, 187, 1977.
277. **Yano, Y.**, Development of positron emitting radionuclides for imaging with improved positron detectors, in *Medical Radionuclide Imaging*, Vol. 2, IAEA, Vienna, 1976, 33.
278. **Robinson, G. D.**, Copper-62: a short-lived, generator produced, positron emitting radionuclide for radiopharmaceuticals, *J. Nucl. Med.*, 17, 559, 1976.
279. **Lewis, M. B.**, Nuclear data sheets, for A = 68, *Nucl. Data Sheets*, 14, 155, 1975.
280. **Greene, M. W. and Tucker, W. D.**, An improved gallium-68 Cow, *Int. J. Appl. Radiat. Isotopes*, 12, 62, 1961.
281. **Welch, M. J. and Wagner, S. J.**, Preparation of positron emitting radiopharmaceuticals, in *Recent advances in Nuclear Medicine*, Vol. 5, Lawrence, J. H. and Budinger, T. F., Eds., Grune & Stratton, New York, 1971, 61.
282. **Kopecky, P. and Mudrova, B.**, ^{68}Ge- ^{68}Ga generator for the production of ^{68}Ga in an Ionic form, *Int. J. Appl. Radiat. Isotopes*, 25, 263, 1974.
283. **Arino, H., Skralsa, W. J., and Kramer, H. H.**, A new ^{68}Ge- ^{68}Ga radioisotope generator system, *Int. J. Appl. Radiat. Isotopes*, 29, 117, 1978.
284. **Neirinckx, R. D. and Davis, M. A.**, Potential column chromatography for ionic Ga-68: organic ion exchangers as chromatographic supports, *J. Nucl. Med.*, 21, 81, 1980.
285. **Loc'h, C., Mazière, B., and Comar, D.**, A new generator for ionic gallium-68, *J. Nucl. Med.*, 21, 171, 1980.
286. **Lewis, R. E. and Camin, L. L.**, Germanium-68/gallium-68 generator for the one step elution of ionic gallium-68, *J. Label. Comp. Radiopharm.*, 18, 164, 1981.
287. **Layne, W. W. and Davis, M. A.**, Development of a gallium-68 generator on alumina, *J. Nucl. Med.*, 21, 85, 1980.
288. **Schumacher, J. and Maier-Borst, W.**, A new ^{68}Ge/^{68}Ga radioisotope generator system for production of ^{68}Ga in dilute HCl, in *Medical Radionuclide Imaging*, IAEA, Vienna, 1981.

289. **Schumacher, J., Maier-Borst, W., and Wellman, H. N.,** Liver and kidney imaging with Ga-68-labeled dihydroxyanthraquinones, *J. Nucl. Med.,* 21, 983, 1980.
290. **Lemming, J. F. and Auble, R. L.,** Nuclear data sheets for A = 82, *Nuclear Data Sheets,* 15, 315, 1975.
291. **Hoop, B., Beh, R. A., Beller, G. A., Brownell, G. L., Burnham, C. A., Hnatowich, D. J., Moore, R. H., Parker, J. A., Roux-lough, P O., Smith, T. W., Budinger, T. F., Chu, P., Yano, Y., Barnes, J. W., Grant, P. M., Ogard, A. E., and O'Brien, H. A.,** Myocardial positron scintigraphy with short-lived ^{82}Rb, *IEEE Trans. Nucl. Sci.,* NS-23, 584, 1976.
292. **Yano, Y. and Anger, H. O.,** Visualization of heart and kidneys in animals with ultrashort-lived ^{82}Rb and the positron scintillation camera, *J. Nucl. Med.,* 9, 412, 1968.
293. **Yano, Y., Budinger, T. F., Chu, P., Grant, P. M., Ogard, A. E., O'Brien, H. A., and Hoop, B.,** Evaluation of ^{82}Rb generators for imaging studies, *J. Nucl. Med.,* 17, 536, 1976.
294. **Budinger, T. F., Yano, Y., and Hoop, B.,** A comparison of ^{82}Rb and ^{13}NH$_3$ for myocardial perfusion scintigraphy, *J. Nucl. Med.,* 16, 429, 1975.
295. **Grant, P. M., Erdal, B. R., and O'Brien, H. A.,** A ^{82}Sr-^{82}Rb isotope generator for use in nuclear medicine, *J. Nucl. Med.,* 16, 300, 1975.
296. **Bertrand, F. E.,** Nuclear data sheet for A = 122, *Nucl. Data Sheets,* B7, 419, 1972.
297. **Richards, P. and Ku, T. H.,** The ^{122}Xe-^{122}I system: a generator for the 3.62m positron emitter ^{122}I, *Int. J. Appl. Radiat. Isotopes,* 30, 250, 1979.
298. **Lederer, C. M. and Shirley, V. S.,** *Table of Isotopes,* 7th ed., John Wiley & Sons, New York, 1978, 28.
299. **Crouzel, C., Guenard, H., Comar, D., Soussaline, F., Loc'h, C., and Plummer, D.,** A new radioisotope for lung ventilation studies: 19-Neon, *Eur. J. Nucl. Med.,* 5, 431, 1980.
300. **Weast, R. C., Ed.,** *Handbook of Chemistry and Physics,* 49th ed., CRC Press, Boca Raton, Fla., 1968.
301. **Lederer, C. M. and Shirley, V. S.,** *Table of Isotopes,* 7th ed., John Wiley & sons, New York, 1978, 58.
302. **Sahakunda, S. M., Qaim, S. M., and Stöcklin, G.,** Cyclotron production of short-lived ^{30}P, *Int. J. Appl. Radiat. Isotopes,* 30, 3, 1979.
303. **Lederer, C. M. and Shirley, V. S.,** *Table of Isotopes,* 7th ed., John Wiley & Sons, New York, 1978, 84.
304. **Lambrecht, R. M., Hara, T., Gallagher, B. M., Wolf, A. P., Ansari, A., and Atkins, H.,** Cyclotron isotopes and radiopharmaceuticals. XXVIII. Production of potassium-38 for myocardial perfusion studies, *Int. J. Appl. Isotopes,* 29, 667, 1978.
305. **Hurst, D G. and Walke, H.,** The induced radioactivity of potassium, *Phys. Rev.,* 51, 1033, 1937.
306. **Tilbury, R. S., Myers, W. G., Chandra, R., Dahl, J. R., and Lee, R.,** Production of 7.6 minute potassium-38 for medical use, *J. Nucl. Med.,* 21, 867, 1980.
307. **Myers, W. G.,** Radiopotassium-38 for in vivo studies of dynamic processes, *J. Nucl. Med.,* 14, 359, 1973.
308. **Hevesy, G. S.,** The application of radioactive indicators in biochemistry, *J. Chem. Soc.,* 1618, 1951.
309. **Beene, J. R.,** Nuclear data sheet for A = 52, *Nucl. Data Sheets,* 25, 235, 1978.
310. **Thakur, M. L., Nunn, A. D., and Waters, S. L.,** Iron-52: improving its recovery from cyclotron targets. *Int. J. Appl. Radiat. Isotopes,* 22, 481, 1971.
311. **Yano, Y. and Anger, H. O.,** Production and chemical processing of ^{52}Fe for medical use, *Int. J. Appl. Radiat. Isotopes,* 16, 153, 1965.
312. **Murakami, J., Akiha, F., and Ezawa, O.,** Comparative studies on the advantage of ^{4}He or ^{3}He particles bombardment for the production of ^{52}Fe and ^{123}I for medical use, in *Radiopharmaceuticals and Labelled Compounds,* Vol. 1, IAEA, Vienna, 1973, 257.
313. **Knospe, W. H., Rayudu, V. M. S., Cardello, M., Friedman, A. M., and Fordham, E. W.,** Bone marrow scanning with 52-Iron, *Cancer,* 37, 1432, 1976.
314. **Lederer, C. M. and Shirley, V. S.,** *Table of Isotopes,* 7th ed., John Wiley & Sons, New York, 1978, 289.
315. **Jones, T. and Matthews, C. M. E.,** Tissue perfusion measured using the ratio of ^{81}Rb to ^{81m}Kr, incorporated in the tissue, *Nature (London),* 230, 119, 1971.
316. **Harper, P. V., Rich, B., Lathrop, K. A., and Mock, B.,** Production and use of ^{81}Rb-^{81}Kr for clinical tissue perfusion measurements with the Anger Camera, in *Dynamic Studies with Radioisotopes in Medicine,* 1974, Vol. 2, IAEA, Vienna, 1975, 133.
317. **Li-Scholz, A. and Bakhru, H.,** Studies in the decay of 4.7 h ^{81}Rb., *Phys. Rev.,* 168, 1244, 1968.
318. **Schneider, R. J. and Goldberg, C. J.,** Production of rubidium-81 by the reaction ^{85}Rb(p,5n)^{81}Sr and decay of ^{81}Sr, *Int. J. Appl. Radiat. Isotopes,* 27, 189, 1976.
319. **Idoine, J. D., Holman, B. L., Jones, A. G., Schneider, R. J., Schroeder, K. L., and Zimmerman, R. E.,** Quantification of flow in a dynamic phantom using ^{81}Rb-^{81m}Kr, and a NaI detector, *J. Nucl. Med.,* 18, 570, 1977.
320. **Peek, N. F., Hegedus, F., DeNardo, G. L., Lagunas-Solar, M., and Berman, D. S.,** Production and characteristics of ^{81}Rb for myocardial studies, *Int. J. Nucl. Med. Biol.,* 5, 273, 1978.

321. **Horigucki, T., Noma, H., Yoshizawa, J., Takemi, H., Hasai, H., and Kiso, Y.**, Excitation functions of proton induced nuclear reactions on ^{85}Rb, *Int. J. Appl. Radiat. Isotopes*, 31, 141, 1980.
322. **van Herk, G., Vaalburg, W., Paans, A. M. J., and Woldring, M. G.**, The preparation of pure ^{81}Rb/^{81m}Kr for dynamic blood flow studies by the ^{81}Rb(p,5n)^{81}Sr and ^{81}Rb(^{4}He,p7n)^{81}Sr reaction, *J. Label. Comp. Radiopharm.*, 13, 247, 1977.
323. **Jones, T., Pettit, J. E., Rhoden, O. G., et al.**, Use of ^{81}Rb-^{81m}Kr ratio for the measurement of spleen blood flow, *J. Nucl. Med.*, 14, 414, 1973.
324. **Lambrecht, R. M., Gallagher, B. M., Wolf, A. P., and Bennet, G. W.**, Cyclotron isotopes and radiopharmaceuticals-XXIX ^{81,82m}Rb for positron emission tomography, *Int. J. Appl. Radiat. Isotopes*, 31, 343, 1980.

Chapter 3

OTHER CYCLOTRON RADIONUCLIDES

Tadashi Nozaki

TABLE OF CONTENTS

I. Radiohalogens 104
   A. Use of Radiohalogens 104
   B. Production of ^{18}F 104
      1. Production Reactions 104
      2. Aqueous Solution of No-Carrier-Added ^{18}F 105
      3. Anhydrous ^{18}F as Precursors for Organic Labeling ... 108
      4. Direct ^{18}F-Labeling of Organic Compounds 109
   C. ^{34m}Cl 110
   D. Radiobromine (^{75}Br, ^{76}Br, and ^{77}Br) 111
      1. Useful Radiobromine Nuclides 111
      2. Production Reactions 111
      3. Target and Chemical Processing 113
      4. Radiobromine Labeling 114
   E. Radioiodine 114
      1. Importance of ^{123}I and Its Production Reactions 114
      2. ^{123}I Production by Direct Route 115
      3. ^{123}I Production Via ^{123}Xe 116
      4. Use of ^{123}I 117
   F. Astatine 117

II. Middle and Relatively Long-Lived Cyclotron Radionuclides Used in Biology and Medicine 118
   A. Useful Nuclides 118
   B. ^{67}Ga, ^{111}In, and ^{201}Tl 119
   C. ^{73}Se 119
   D. ^{52}Fe 120
   E. ^{28}Mg and ^{43}K 120
   F. ^{203}Pb and ^{237}Pu 121
   G. Other Nuclides 121

References 121

I. RADIOHALOGENS

A. Use of Radiohalogens

Chlorine, bromine, and iodine are familiar in organic chemistry, and radiohalogen labeling has been studied extensively for the past 30 years. Formerly reactor-produced ^{36}Cl, ^{82}Br, and ^{131}I were used almost exclusively. Recently cyclotron-produced ^{18}F, ^{77}Br, and ^{123}I have become more and more important especially in nuclear medicine.

Abundances of F, Cl, Br, and I in earth's crust are 630 ppm, 300 ppm, 2 ppm, and 500 ppb, respectively, and in sea water are 1.3 ppm, 1.9%, 65 ppm, and 60 ppb, respectively. Only quite limited varieties and quantities of organic halogen compounds exist in normal nature, but the nature at present is contaminated with various manmade organic halogen compounds. Iodine is the least abundant of the four halogens, but is concentrated in sea weeds and thyroid gland as organic compounds.

In biology and medicine, radiohalogens are used both in ionic and organic forms. Radiohalogen-labeled organic compounds are used for the tracing of the two different kinds of compounds: (1) the halogen compounds themselves, and (2) nonhalogens, usually naturally occurring compounds under the assumption that the halogen introduction would have resulted in no notable change in their behavior under examination. The halogen labeling, in the latter case, is a substitute for the labeling of the natural compound or its related compound with a radionuclide of its component element. In H, C, N, and O, there is no γ-ray or positron emitter nuclide with half-life longer than 30 m. In order for the above assumption to be valid, it is necessary for the carbon-halogen bond to be stable enough during the tracer experiment and for the halogen labeling not to cause any marked change of molecular shape and size. As is well known, carbon-halogen bond energy decreases with atomic number of the halogen, and the difference in size and mass between hydrogen and halogen increases with it. Hence, fluorine is the most favorable halogen in this use, with the order of preference being F > Cl > Br > I. However, chemical stability, especially stability in vivo, is markedly dependent on the position of the halogen atom in the molecule.

Halogen substitution is known to sometimes change considerably the metabolic path of biomolecules. This change often offers interesting subjects of tracer study and can also be utilized in nuclear medicine. For example, monofluoracetic acid is highly poisonous, 5-fluorouracil is a cancer suppressing pharmaceutical, and ^{18}F-labeled 2-deoxy-2-fluoroglucose is useful in brain metabolism study. Also, 6-iodomethyl-19-nor derivative of cholesterol shows much higher adrenal concentration than cholesterol, and is used for adrenal imaging in radioiodine labeled state.

As for the commercial availability, ^{123}I is now steadily supplied from various producers in the world in spite of its rather limited life. In some fortunate cases, ^{77}Br and ^{18}F can be purchased. In the preparation of ^{18}F-labeled compounds, ^{18}F should almost always be produced on site. In an in-house radionuclide production, various conditions, such as nuclear reaction, target material, and chemical separation, should be selected mainly on the consideration of (1) the characteristics of the available cyclotron and (2) required purity and physical and chemical state of the product.

B. Production of ^{18}F

1. Production Reactions

Natural fluorine consists only of ^{19}F, and nonradioactive tracer nuclide does not exist in fluorine. Fluorine-18 (110 m half-life; β^+ decay; 0.635 MeV max. β^+ energy; no γ-ray; 9.5×10^7 Ci/g specific activity in carrier-free state) is the longest-lived radiofluorine, though its life is rather short for many applications in which ^{18}F should first be introduced in an organic compound. The low β^+ energy makes this nuclide ideal for the positron-emitter localization measurement.

Table 1
NUCLEAR REACTIONS FOR THE PRODUCTION OF ^{18}F

Bombardment	Predominate reaction	Q-value (MeV)	Threshold energy (MeV)
O + t	^{16}O(t,n)^{18}F	+ 1.270	0
O + ^{3}He	^{16}O(^{3}He,p)^{18}F	+ 2.003	0
	^{16}O(^{3}He,n)^{18}Ne → ^{18}F	− 3.196	3.795
O + α	^{16}O(α,pn)^{18}F	− 18.544	23.180
	^{16}O(α,2n)^{18}Ne → ^{18}F	− 23.773	29.716
Ne + d	^{20}Ne(d,α)^{18}F	+ 2.796	0
Ne + ^{3}He	^{20}Ne(^{3}He,αp)^{18}F	− 2.697	3.102
	^{20}Ne(^{3}He,αn)^{18}Ne → ^{18}F	− 7.926	9.115
^{18}O + p (enriched)	^{18}O(p,n)^{18}F	− 2.436	2.571

Once ^{18}F was produced by reactor irradiation via the two step reactions: ^{6}Li(n,α)t, ^{16}O(t,n)^{18}F. Cyclotron production is now clearly preferred in yield, simplicity, and physical and chemical state of the product. As shown in Table 1, ^{18}F can be produced by various charged particle reactions on oxygen and neon. The excitation functions for these reactions have been known reliably.[1-3] Their thick target yield curves are shown in Figures 1 to 3. In the ^{3}He and α-particle reactions, the path via ^{18}Ne also contributes to the ^{18}F formation. This contribution, however, is regarded as minute, because the Q-value for this path is more negative and ^{18}Ne is an even-even nuclide.[4] Recent measurement for the Ne + ^{3}He → ^{18}F reaction has proven it.[5]

Since tritium acceleration and bombardment always involve various problems concerning tritium contamination, the O + t → ^{18}F reaction is not used for production purpose. In the proton bombardment of ^{18}O, the enriched ^{18}O target should usually be recovered. The recovery is easy for H$_2$^{18}O but rather difficult for ^{18}O$_2$ target, and always some loss of ^{18}O is inevitable. This reaction is actually valuable when the available cyclotron can accelerate no other particles than protons to enough energy for ^{18}F production.

2. Aqueous Solution of No-Carrier-Added ^{18}F

No carrier-added ^{18}F as fluoride in water can be produced by the ^{16}O(^{3}He,p)^{18}F, ^{16}O(α,pn)^{18}F or ^{18}O(p,n)^{18}F reaction, with the target being water in a titanium container. This production is very easy and efficient. Titanium foil is used as the beam-entrance window. A typical target assembly is shown in Figure 4[6], though there are many modified designs. Such apparatus are fabricated by some cyclotron producers as attachments of nuclide-production cyclotrons.

Titanium is preferred as the target container, because it is highly resistive to corrosion in an oxidizing medium. Water under being bombarded is decomposed into H$_2$, O$_2$, and H$_2$O$_2$, etc. and becomes oxidizing. Aluminum and some other metals are corroded considerably resulting in a complex formation of the ^{18}F. For the volume of the target water, several milliliters are enough. The formation of H$_2$ and O$_2$ elevates the pressure when the system is completely closed. Thus, a catalyzer column for the recombination of H$_2$ and O$_2$ is attached to the target container to make it an open system. Water formed by the recombination, however, deteriorates the activity of the catalyzer, covering its surface. The catalyzer seems not to be absolutely necessary, but a good cooling of the target water as well as the evolved gas is essential to keep the ^{18}F in the system. The filling and withdrawal of the target water can be carried out automatically by the application of air pressure.

No danger of the beam-window breakdown by heat evolution needs to be considered, because the window is always cooled by the target water itself. The bombardment with a maximum available energy is regarded as often unprofitable, because (1) the formation of

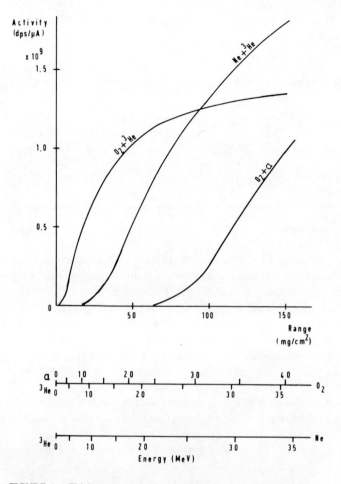

FIGURE 1. Thick target saturation activities for the reactions of ^{3}He-particle on oxygen, ^{3}He-particle on neon, and α-particle on oxygen to give ^{18}F.

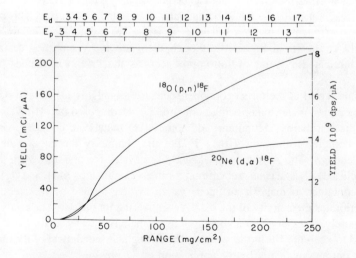

FIGURE 2. Thick target saturation activities for the reactions of deuteron on neon and proton on ^{18}O to give ^{18}F.

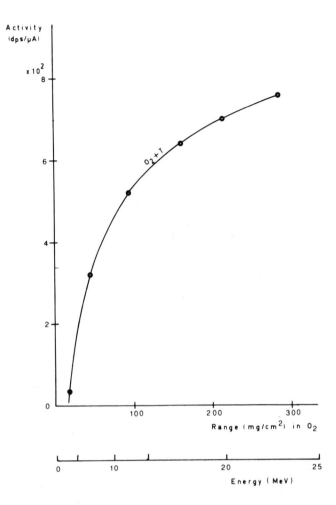

FIGURE 3. Thick target saturation activity for the reaction of triton on oxygen to give ^{18}F.

impurities generally increases with energy, (2) the rate of heat evolution and radiation decomposition increases with it, and (3) the thick target yield of ^{18}F does not increase steeply with it for energy regions far above the maximum of the excitation function. The product ^{18}F contains ^{11}C, ^{13}N, and ^{7}Be formed from the target water itself and also ^{48}V, ^{48}Cr, etc., formed in the titanium window foil and pushed into the water by nuclear recoil. Their activities, however, are too low to be paid any serious attention, unless the ^{18}F is used immediately after the bombardment or many hours after it. Usually over 100 mCi of ^{18}F is produced in one batch, as is understood from Figure 1.

The ^{20}Ne(d,α)^{18}F reaction is also used for the production of aqueous ^{18}F. The ^{18}F formed in neon is readily adsorbed on the inside wall of the bombardment box and can then be washed out into water or an aqueous solution. An automated apparatus has been used for this purpose.[7]

The aqueous no-carrier-added ^{18}F was once widely used for bone scanning, but has been replaced by ^{99m}Tc polyphosphates. Studies have been made for the preparation of various ^{18}F-labeled organic compounds from aqueous ^{18}F both with and without carrier addition.[8,9] This ^{18}F is also expected to be useful in various fields of biology and agricultural research, where the behavior of a trace quantity of fluorine has often still remained obscure.

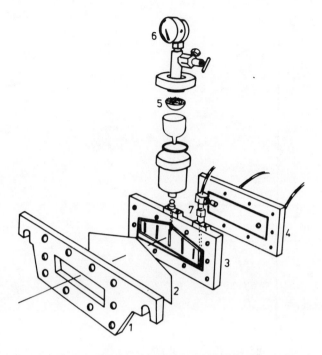

FIGURE 4. A target for production of no-carrier-added ^{18}F in water. (1) Target mounting plate, (2) foil window, (3) target plate with target water cavity, (4) target cooling block with "O" ring, (5) catalyst in container, (6) pressure gauge, (7) tube and valve for filling/emptying cavity with target water.

3. Anhydrous ^{18}F as Precursors for Organic Labeling

Anhydrous forms of fluorine are often needed in the preparation of organic fluorine compounds. The deuteron bombardment of neon is preferred for the ^{18}F production, because it gives a high yield without formation of any oxidizing substance. The bombardment of O_2 gives O_3. As the precursor of ^{18}F labeling, HF, F_2, CF_3OF, BF_3, SF_4, NOF, ClF, KF, CsF, AgF, SbF_3, $NaBF_4$, etc. can be used.

The ^{18}F formed in neon is so readily adsorbed on almost all substances that many techniques have been tried and used for its withdrawal from the target box.[10] These techniques can be grouped into the following three categories. (1) Target neon is continuously circulated through a tube in which a substance to be labeled is supported on a filter or fibers. The solid ^{18}F-labeling precursors can thus be prepared. Also, no-carrier-added K^{18}F on KOH and Cs^{18}F on CsOH are obtained by this way and used for the preparation of some no-carrier-added ^{18}F organic compounds.[11] (2) The ^{18}F is left to be adsorbed in the target box made of a suitable substance, and is then eluted with H_2 under heating to give no-carrier-added H^{18}F or with a nonaqueous solution or suspension of a fluorinating agent under shaking. (3) Neon containing a gaseous fluorinating agent is bombarded under conditions selected for minimizing the ^{18}F loss. This method is successfully used for the preparation of $^{18}F_2$.[12]

Efforts have been made to produce anhydrous H^{18}F efficiently.[13,14] Various conditions in its production by Method 2 have been studied in order to increase the elution efficiency.[14] Copper is shown to be a suitable substance for the target box. No-carrier-added H^{18}F is extracted from the target by purging with 10% H_2 in He while heating the target to 300°C. The yield is reported to be reproducibly 10 mCi/μAh (at the end of saturation bombardment) or 75% of the total ^{18}F formed in the target. The H^{18}F can be caught on a desired ^{18}F-labeling precursors mainly by nuclidic exchange.

By the use of $^{18}F_2$, hydrogen atom bound to a double-bonded carbon atom can be substituted by ^{18}F:

$$-CH=CH- \xrightarrow[\text{or }^{18}F_2-HQ]{^{18}F_2} \left\{ \begin{array}{l} -CH^{18}F-CH^{18}F- \\ -CH^{18}F-CHQ- \end{array} \right\} \xrightarrow{\Delta} -C^{18}F=CH-$$

$$(Q = HO, CH_3CO_2)$$

Via this route, ^{18}F-labeled 2-deoxy-2-fluoroglucose and 5-fluorouracil are synthesized.[15,16] Carrier-free ^{18}F is of so high a specific activity that it is extremely difficult theoretically to obtain any substance with its each molecule containing two or more carrier-free ^{18}F atoms. In the production of ^{18}F-labeled F_2, thus, the addition of some carrier is necessary, though the final ^{18}F organic compounds are usually required in as high a specific activity as possible. Since $^{18}F_2$ is effectively used now, its production is described here in some detail.

As the target gas, neon containing a small amount (e.g., 0.1 vol%) of F_2 is used. Care should be taken to the purity of the gas mixture, because impurity N_2, CO_2 or CF_4, when present over 0.1%, results in the formation of an unacceptable level of $N^{18}F_3$ or $C^{18}F_4$ at the expense of the $^{18}F_2$.[17] The gas should be filled at an elevated pressure. The amount of ^{18}F that can be removed from the target is shown to increase with both the target pressure and carrier concentration. Nickel is a suitable substance for the target box and inner side of the beam window. The window should have some thickness in order to withstand the inside pressure and heat evolved in it during the bombardment. A sandwich of nickel and aluminum foils is preferred as the window, because aluminum is a good thermal conductor. It is essential to passivate the inner surfaces of the target box in order to minimize the trapping of ^{18}F by them. For the passivation, dry F_2 is introduced into the box (50 to 200 Torr), which is then heated to dull red (400 to 600°C). In routine use, the passivation should be repeated fairly frequently.

A representative target design is shown in Figure 5.[12] There are two cooling systems, one for the target box and the other for the beam window. As yet the possibility of the beam window rupture needs to be taken into account, and the beam should be spread uniformly or caused to scan over a certain area. As is known from Figure 2, the energy of deuteron is preferred to be from 7 to 15 MeV after it has passed through the window. In the use of a low-energy cyclotron, the window should be made thinner and cooled efficiently by, e.g., a cold helium stream. The target length should be determined from the deuteron energy and the gas pressure by the use of the range-energy curve. The target gas, however, loses its density in the beam path due to the temperature rise and plasma formation, and this should be taken into account in the design of target box. Practically, it is essential (1) to set the target assembly at a fixed position together with the inlet and outlet equipments for the target gas and passivation gas and (2) not to expose the inside of this system to air when it is not used. Those who are going to produce $^{18}F_2$ by themselves are asked to read more detailed descriptions in Reference 12.

A typical example for $^{18}F_2$ production is given here:[12] machine energy, 23 MeV; deuteron energy on target, 14.0 MeV; target neon pressure, 25.8 atm; carrier F_2 concentration, 0.1%; beam current, 15 μA; dose on target, 30 μA h; theoretical yield, 655 mCi; recovered yield, 367 mCi.

4. Direct ^{18}F-Labeling of Organic Compounds

When an organic gas, such as a chloromethane or fluorochloromethane, is added to the target neon, ^{18}F-labeled product gases are obtained.[19] When the inside of a neon target box is coated with a solid substance, such as uracil, it can be labeled with no-carrier-added ^{18}F.[20] The yield of the labeled compounds usually does not increase linearly with the total deuteron dose, but shows a saturation or even decrease with it. Direct labeling in general gives a

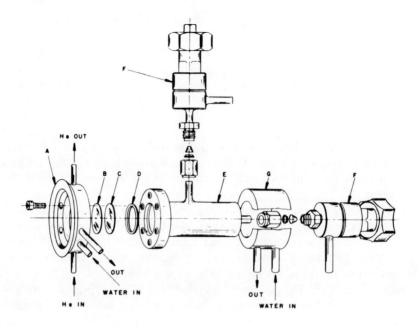

FIGURE 5. A target for production of ^{18}F-F2. (A) Front cooling flange constructed of brass or Al cut to fit Marmon flange (o.d. 9.68 cm). (B) Aluminum degrader foil; 0.18 mm. (C) Nickel foil; 0.025 mm. (D) Nickel pressure ring "O" or "C", plated with gold or silver (0.318 × 2.9 cm i.d.). (E) Target body; nickel pipe 2.5 cm i.d., 10 cm in length; "O" ring groove dimensions (0.25 cm deep, 2.7 cm i.d., 3.5 cm o.d.). (F) Hoke diaphragm valves constructed of monel; KEL-F insulators connect stainless steel transport line to these valves. (G) Brass cooling jacket; sliding fit with rubber "O" rings for sealing.

mixture of labeled compounds, and their separation is tedious and difficult. Exceptionally, however, the bombardment of neon containing small amounts of COF_2-F_2 or CF_4-F_2 gives $C^{18}F_4$ in high yield and purity.[16,21]

C. ^{34m}Cl

Chlorine-34m (half-life: 32 m; decay: β^+ (53%) and IT (47%); max. β^+ energy: 2.5 MeV; γ-ray energy: 0.145 MeV, 2.12 MeV, etc.; specific activity in carrier-free state: 1.7×10^8 Ci/g) is the sole neutron-deficient radiochlorine available in biomedical tracer study, though not so useful as ^{18}F, ^{77}Br, or ^{123}I. In the IT route it decays into ^{34}Cl (half-life: 1.6 s; decay: β^+; max. β^+ energy: 4.5 MeV; no γ-ray). Their positron energies are too high for positron-emitter localization, and ^{34m}Cl emits hard γ-rays. For the production of ^{34m}Cl, bombardments of P, S, and Cl with p, d, ^{3}He, and α-particles were studied. The ^{3}He and α-particle reactions on sulfur are the most efficient for the production of no-carrier-added ^{34m}Cl.[22] Excitation functions have been reported for the ^{32}S(^{3}He,p)^{34m}Cl and ^{35}Cl(p,pn)^{34m}Cl reactions, the latter giving ^{34m}Cl in target chlorine.[23,24] Thick target yields of 0.086 and 1.02 mCi/μA m are reported for the former reaction with 25 MeV ^{3}He and for the latter reaction with 24 MeV proton, respectively. The targetry and chemical treatment of the product ^{34m}Cl are still left to be studied further.

The reactions ^{40}Ar(p,2p)^{39}Cl and ^{40}A(γ,p)^{39}Cl give no-carrier-added ^{39}Cl, which is a β^- and γ-ray emitter with 56 m half-life. Although this nuclide can also be used as a tracer, it cannot be produced efficiently. By reactor irradiation, ^{36}Cl (3×10^5 y, β^-, no γ) and ^{38}Cl (37 m, energetic β^- and γ-rays) are produced, and ^{36}Cl has been widely used in organic chemistry and biology.

Table 2
RADIOBROMINES WITH T > 1 H AND THEIR PARENT RADIOKRYPTONS

Nuclide	Half-life	Decay mode	Particle energy (MeV)	Main γ-ray energy (MeV)	Specific activity (Ci/g)
^{75}Br	95 m	β$^+$(75%), EC(25%)	1.7	0.286	2.6×10^7
^{76}Br	16 h	β$^+$(57%), EC(43%)	3.4	0.559	2.6×10^6
^{77}Br	56 h	EC		0.239	7.3×10^5
^{80m}Br	4.4 h	IT		0.037	8.9×10^6
^{82}Br	35 h	β$^-$	0.44	0.776	1.1×10^6
^{83}Br	2.4 h	β$^-$	0.93	(0.53)	1.6×10^7
^{75}Kr	4.2 m	β$^+$, EC	3.2	0.132	6.0×10^8
^{76}Kr	14.8 h	EC		0.252	2.8×10^6
^{77}Kr	75 m	β$^+$(80%), EC(20%)	1.86	0.130	3.2×10^7

Table 3
NUCLEAR REACTIONS FOR RADIOBROMINE PRODUCTION

Radiobromine	Reaction	Target nuclide abundance (%)	Q-value (MeV)	Main ^{76}Br-formation reaction
^{75}Br	^{75}As(^{3}He,3n)^{75}Br	100	−12.9	^{75}As(^{3}He,2n)^{76}Br
	^{76}Se(p,2n)^{75}Br (Enriched target)	9.0	−14.7	^{76}Se(p,n)^{76}Br
	^{75}As(α,2n)^{77}Br	100	−13.5	^{75}As(α,3n)^{76}Br
	Se + p → ^{77}Br			^{77}Se(p,2n)^{76}Br
^{77}Br	Mainly ^{78}Se(p,2n)^{77}Br	23.5	−12.6	^{78}Se(p,3n)^{76}Br
	^{79}Br(p,3n)^{77}Kr → ^{77}Br	50.5	−22.6	^{79}Br(p,4n)^{76}Kr → ^{76}Br
	^{79}Br(d,4n)^{77}Kr → ^{77}Br	50.5	−24.8	^{79}Br(d,5n)^{76}Kr → ^{76}Br

D. Radiobromine (^{75}Br, ^{76}Br, and ^{77}Br)

1. Useful Radiobromine Nuclides

Now two cyclotron-made radiobromines, ^{75}Br and ^{77}Br, are regarded as more useful than reactor-made ^{82}Br. Radiobromines with half-lives longer than 1 h are listed in Table 2, together with related radiokryptons. Many organic molecules can be labeled with radiobromines much more easily than with ^{18}F. Also a C-Br bond is more stable than the corresponding C-I bond. ^{75}Br is a suitable nuclide for positron-emitter localization, though it has not as yet been widely used. ^{77}Br emits γ-ray of a suitable energy for dianostic purposes. Since ^{77}Br is the longest-lived radiobromine, it is also appreciated in long-term tracer experiments.

2. Production Reactions

Nuclear reactions useful for the production of radiobromines are listed in Table 3. The excitation functions have been measured for all these reactions and for simultaneous formation of ^{76}Br.[25-30] Some of the excitation functions are shown in Figures 6 to 8, and the thick target yields for them are given in Table 4.[25] In the production of ^{77}Br and ^{76}Br via their parent radiokryptons, the kryptons are first separated from the target bromine and then left to decay into the radiobromines. Of the four ^{75}Br formation reactions, the ^{76}Se(p,2n)^{75}Br reaction on enriched ^{76}Se gives the highest yield of nearly 100 mCi/μA h with 1.4% ^{76}Br contamination when proton energy range between 28 and 22 MeV is used.[29] In the ^{75}As(^{3}He,3n)^{75}Br reaction, use of ^{3}He energy range between 36 and 25 MeV is shown to give 36 mCi/μA h of ^{75}Br accompanied with minute amounts of ^{76}Br and ^{77}Br.[30]

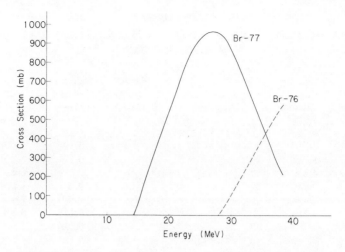

FIGURE 6. Excitation functions for the ^{75}As(α,2n)^{77}Br and ^{75}As(α,3n)^{76}Br reactions.

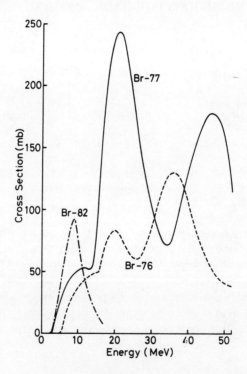

FIGURE 7. Excitation functions for the Se + p → ^{77}Br, ^{76}Br, ^{82}Br reactions. Target: natural selenium.

For practical ^{77}Br production the ^{75}As(α,2n)^{77}Br reaction has been used most commonly, though the proton reactions give higher yields. The nuclidic purity of ^{77}Br becomes higher with time, because other radiobromines decay more rapidly. In the ^{76}Br and ^{77}Br productions via the radiokryptons, mutual contamination can be suppressed by proper selection of the time between the separation of the kryptons from the target and that of the radiobromines from the kryptons. For actual production, the nuclear reaction, the incident energy, and target thickness should be selected on consideration of the characteristics of the available cyclotron, the required purity of the product, as well as various practical problems concerning targetry and product processing.

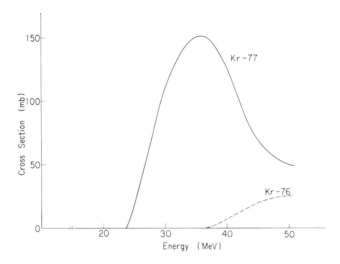

FIGURE 8. Excitation functions for the Br + p → ^{77}Kr and ^{76}Kr reactions. Target: natural bromine.

Table 4
THICK TARGET SATURATION ACTIVITY (10^9DPS μA^{-1}) OF ^{77}BR, ^{76}BR AND ^{82}BR. (TARGET: ELEMENT OF NATURAL NUCLIDIC COMPOSITION)

	Reaction	10	15	20	25	30	35	40	45	50
Se + p	↗ ^{77}Br	0.43	1.1	3.9	7.7	10.6	12.4	15.3	19.5	25.1
	→ ^{76}Br	0.17	0.78	1.8	3.0	4.5	7.2	10.4	12.5	13.9
	↘ ^{82}Br	0.66	1.03	1.1	1.1					
Br + p	↗ ^{77}Kr → ^{77}Br	0	0	0	0.18	2.0	5.4	8.7	11.1	12.8
	↘ ^{76}Kr → ^{76}Br	0	0	0	0	0	0	0.06	0.49	1.2
As + α	↗ ^{77}Br	0	0	0.32	1.0	1.9	2.5	2.8		
	↘ ^{76}Br	0	0	0	0	0.03	0.31	0.91		

3. Target and Chemical Processing

Since all target elements in Table 3 are rather volatile, considerable efforts have been made in the targetry for radiobromine production. When the α-particle beam can be spread over a large target area, As$_2$O$_3$ is the most effective form of arsenic target.[31] For a high beam intensity, Cu-As alloy is used.[32] Now GaAs is produced for semiconductor use, and this is also a good substance for arsenic target. Various chemical forms of selenium are used as selenium target, such as elementary selenium, PbSe, Al$_2$Se$_3$, and SeO$_2$-Na$_2$O-B$_2$O$_3$, though they are less resistive to high beam intensity than the arsenic targets.[25,33] Any target should be cooled efficiently and often be covered with a metal foil in order to retain the radiobromine.

For the separation of radiobromine from an arsenic or selenium target, the following three methods have been used: (1) distillation from a solution,[25,31,34] (2) volatilization by fusion of the target,[32] and (3) coprecipitation with AgCl followed by the dissolution of the precipitate in NH$_3$ water and subsequent removal of cations by ion exchange.[33] Each method has its advantage, but the volatilization by radio-frequency heating seems to be well suited for a large-scale production. Possible contamination of nonradioactive bromine from reagents and laboratory atmosphere should always be kept in mind in chemical processing.[25] A method is shown to obtain no-carrier-added ^{77}Br in water containing no other solute.[25] Bromine in this state is gradually lost when the solution is concentrated by evaporation.

Table 5
NEUTRON-DEFICIENT RADIOIODINES WITH T > 1 H AND THEIR PARENT RADIOKRYPTONS

Nuclide	Half-life	Decay mode	Particle energy (MeV)	Main γ-ray energy (MeV)	Specific activity (Ci/g)
^{120}I	1.4 h	EC(54%), β$^+$(46%)	4.6	0.56	1.9×10^7
^{121}I	2.1 h	EC(94%), β$^+$(6%)	1.2	0.213	1.2×10^7
^{123}I	13 h	EC		0.159	1.9×10^6
^{124}I	4.2 d	EC(75%), β$^+$(25%)	2.1	0.603	2.5×10^5
^{125}I	60 d	EC		0.035	1.7×10^4
^{126}I	13 d	EC(53%), β$^-$(46%), β$^+$	0.87	0.67	8.0×10^4
^{120}Xe	40 m	EC(97%), β$^+$(3%)		0.073	3.9×10^7
^{121}Xe	39 m	EC(92%), β$^+$(8%)	2.8	0.253	4.0×10^7
^{123}Xe	2.1 h	EC(87%), β$^+$(13%)	1.5	0.149	1.2×10^7
^{125}Xe	17 h	EC		0.188	1.5×10^6

The ^{77}Br production via ^{77}Kr is of the same principle as the ^{123}I production via ^{123}Xe, which is described in Section I.E.3. Some difficulty, however, is encountered in continuous complete trapping by simple cooling, because krypton is of a higher boiling point than xenon. Simple techniques are described for a small-scale production of ^{77}Br via ^{77}Kr and its use in organic labeling.[35] Routine large-scale production by this route has not as yet been reported.

4. Radiobromine Labeling

Many methods have been used for ^{77}Br labeling of organic compounds, including those developed for ^{82}Br labeling.[36] Aromatic rings of some compounds are labeled with no-carrier-added radiobromine in water with the aid of Chloramine T or H_2O_2 and an enzyme. The tosylate and mesylate process as well as iodine-to-bromine interchange can also be used for the no-carrier-added labeling. Isotopic exchange often offers the best labeling method, when the presence of a minute amount of carrier is tolerable. The decay of ^{77}Kr into ^{77}Br can be utilized for the excitation labeling with ^{77}Br, but its yield usually is considerably poorer than in the excitation labeling with ^{123}I.

E. Radioiodine

1. Importance of ^{123}I and Its Production Reactions

A large amount of radioiodine is produced and used mainly in nuclear medicine. Neutron-deficient radioiodine with half-lives longer than 1 hr are listed in Table 5, together with related radioxenons. Formerly reactor-made ^{131}I was used almost exclusively. Now most part of ^{131}I is going to be replaced by cyclotron-made ^{123}I, because the latter gives much lower internal radiation dose to the patient and because its γ-ray energy is ideally suited to many gamma cameras. For administration into a patient ^{123}I should be produced as free from ^{124}I as possible, because ^{124}I gives much higher radiation dose. The ^{123}I/^{124}I activity ratio decreases with time. An example of radiation dose in thyroid diagnosis is shown in Figure 9 for various sources of radioiodine as a function of time-lapse between the end of bombardment and the administration.[37]

Any other cyclotron-made radioiodine is seldom used. Although ^{121}I has good characteristics for use in nuclear medicine, it decays into ^{121}Te (17 d, EC, γ-rays). At present, ^{125}I is produced in large amounts and used in radioimmunoassay. Although this nuclide can be produced by various charged particle reactions, the ^{124}Xe (n, γ)^{125}Xe $\xrightarrow[17\ \text{hr}]{EC}$ ^{125}I reaction induced by reactor irradiation is preferred for large-scale production.

There are two practical different routes in the production of ^{123}I: (1) direct production, mainly by the ^{124}Te(p,2n)^{123}I reaction on an enriched ^{124}Te, and (2) production via ^{123}Xe,

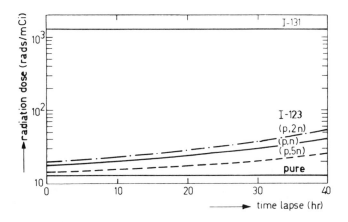

FIGURE 9. Radiation dose in thyroid diagnosis for various sources of iodine as a function of time-lapse between EOB and administration (per mCi of iodine administered).

mainly by the ^{127}I(p,5n)^{123}Xe $\xrightarrow[2.1\ hr]{EC}$ ^{123}I reaction. In principle, ^{124}I-free product can be obtained only by the ^{123}Xe route. It is sometimes necessary for the users of commercial ^{123}I to know its production route. Excitation functions or yield curves were measured for various reactions available for ^{123}I production. Especially for the ^{124}Te(p,2n)^{123}I and ^{127}I(p,5n)^{123}Xe reactions, the measurements were repeated carefully by various different groups and the results compared.[37-40] Since ^{123}I is now commercially available, only an outline of its production is given here.

2. ^{123}I Production by Direct Route

Beside the ^{124}Te(p,2n)^{123}I reaction, several reactions have been studied and used in small-scale production.[36] These include ^{123}Te(p,n)^{123}I, ^{122}Te(d,n)^{123}I, ^{123}Sb(^{3}He,3n)^{123}I and ^{121}Sb(α,2n)^{123}I reactions. They are, however, poorer in yield and/or product nuclidic purity than the ^{124}Te(p,2n)^{123}I reaction on a highly enriched target under a selected proton energy range.

Differential yield curves of ^{123}I and ^{124}I in the proton bombardment of enriched ^{124}Te targets are shown in Figure 10.[39] Proton energy of a compact AVF cyclotron is well suited to this production. There are two nuclear reactions giving ^{124}I contamination in the proton reaction on an enriched ^{124}Te, which inevitably contains ^{125}Te: (1) the ^{125}Te(p,2n)^{124}I reaction, which has a similar excitation function as the ^{124}Te(p,2n)^{124}I reaction; and (2) the ^{124}Te(p,n)^{124}I reaction, which has its maximum cross-section at a lower proton energy. (Isotopic composition of natural tellurium: ^{120}Te, 0.091%; ^{122}Te, 2.5%; ^{123}Te, 0.89%; ^{124}Te, 4.6%; ^{125}Te, 7.0%; ^{126}Te, 18.7%; ^{128}Te, 31.9%; ^{130}Te, 34.5%.) In order to minimize the ^{124}I contamination, it is thus essential (1) to use the ^{124}Te target as free from ^{125}Te as possible, and (2) to select the incident proton energy and target thickness. The ^{123}I/^{124}I activity ratio has been studied carefully, though some discrepancy exists among its values obtained by calculation from the yield curves of different groups and by actual measurements.

Target preparation, chemical separation of the ^{123}I, and recovery of the enriched ^{124}Te were studied well,[37,40] and satisfactory procedures have been established for commercial routine production. Elementary tellurium powder pressed into a layer of a suitable thickness is bombarded under an efficient cooling, and the product ^{123}I is separated either by volatilization in dry fusion or by some wet method, such as distillation or sorption on platinum.[37,40] The following is an example for practical routine production:[39] incident proton energy, 27 MeV; target thickness, 225 mg/cm^2; ^{124}Te enrichment, 96.21%; production rate, $\geq$ 10 mCi/μA h; routine productivity, $\geq$ 200 mCi/h; radionuclidic impurity at EOB, ^{124}I 0.78%, ^{125}I 0.01%, and ^{126}I 0.07%.

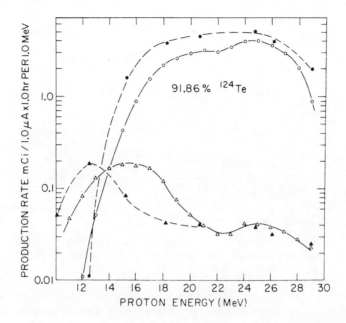

FIGURE 10. A comparison of the production rates of ^{123}I and ^{124}I for the proton reactions on enriched ^{124}Te. Enrichment: 91.86%. ○ ^{123}I by Kondo et al. (Ref. 39); ● ^{123}I by Acerbi et al. (Ref. 38); △ ^{124}I by Kondo et al. (Ref. 39); ▲ ^{124}I by Acerbi et al. (Ref. 38).

3. ^{123}I Production via ^{123}Xe

The production via ^{123}Xe can give ^{124}I-free ^{123}I, because ^{124}Xe is stable. In Figure 11 are shown the excitation functions for the ^{127}I(p,3n)^{125}Xe, ^{127}I(p,5n)^{123}Xe, and ^{127}I(p,7n)^{121}Xe reactions in the latest publication, where the comparison with older results is also given.[41] In this process, the product xenon is usually swept out continuously from the target with the aid of a helium stream, made free from mists and iodine or its compounds in a pre-trap cooled by dry ice or in a suitable column, and then trapped in a vessel by liquid-nitrogen cooling. The ^{123}Xe is left to decay in the vessel until the ^{123}I activity reaches the maximum value, and the ^{123}I is washed out usually with water. As is clear from Figure 11, a high energy proton beam is necessary in this process (over 50 MeV, preferably 80 MeV). Thus, a cyclotron for this use is very expensive. The selections of the incident proton energy and target thickness are necessary, because of the formation of ^{125}Xe-^{125}I and ^{121}Xe-^{121}I-^{121}Te in lower and higher proton energy regions, respectively. The formation of ^{122}Xe need not be considered, because ^{122}I is of only 4 m half life and ^{122}Te is stable.

The production of ^{123}I has also been studied by the use of other reactions, such as deuteron reactions on tellurium,[42] α-particle reactions on antimony, iodine, and tellurium.[43] Also, still higher energy proton reactions (320 to 660 MeV) on CsCl, BaCO$_3$ and La$_2$O$_3$ targets were studied in view of the production of radioxenons.[44,45] The protons are accelerated by machines designed mainly for high energy physics use. Thick target yields of radioxenons are shown to be very high, but the radionuclidic purity probably offers a serious problem.

In the reaction on iodine, physical and chemical states of the target should be selected on consideration of its iodine content, resistivity to radiation decomposition and reactivity to container substances, as well as the ease of the xenon extraction. Various materials have been tested as the target, such as powders or saturated solutions of KI and NaI and CH$_2$I$_2$ containing I$_2$.[36,46,47] The solid target substance is pressed to a porous mass, through which a helium stream is passed to carry the ^{123}Xe. The liquid target substance is usually circulated, and the ^{123}Xe is extracted into helium bubbles passing through it. Also, various equipments were devised for this purpose.[36,46,48]

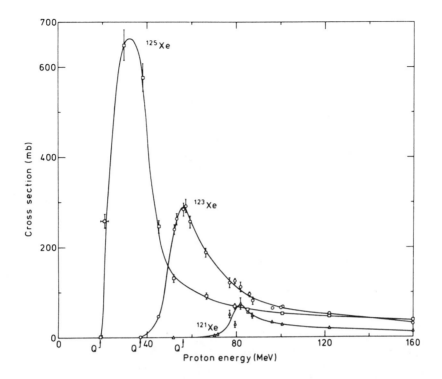

FIGURE 11. Excitation functions for the $^{127}I(p,3n)^{125}Xe$, $^{127}I(p,5n)^{123}Xe$, and $^{127}I(p,7n)^{121}Xe$ reactions.

An example of the production for domestic use is shown here: proton energy range, 60 to 40 MeV; proton flux, 4 µA; target, saturated aqueous solution of NaI (15 mℓ); flow rate of He, 20 mℓ/m; ^{123}Xe purification, dry ice-acetone pre-trap; ^{123}Xe trap, liquid-N_2 cooled glass vessel; bombardment duration, 7 h; time for ^{123}Xe decay after EOB, 4.5 h. Thus, about 130 mCi of ^{123}I is washed out in water from the vessel. In a large-scale routine production, 1 Ci or still much more of ^{123}I is produced each time and distributed to users in wide areas of the world by air transportation.

4. Use of ^{123}I

Iodine is known to have various oxidation states, among which the nuclidic exchange is often complicated. Organic labeling with ^{123}I has been studied fairly well.[36] Aromatic rings of some bio-molecules can be labeled with no-carrier-added ^{123}I by the Chloramine T or enzymatic method, just as with ^{77}Br. Most iodo-compounds are labeled easily and efficiently by nuclidic exchange with ^{123}I iodide. Care should be taken, however, to the relative unstability of the C-I bond. Excitation labeling by the use of the $^{123}Xe \rightarrow {}^{123}I$ decay gives no-carrier-added ^{123}I compounds.

Two longer-lived reactor-made radioiodine, ^{125}I and ^{131}I, are readily available commercially in no-carrier-added state. Except in diagnostic administration, these reactor nuclides are often preferable when used as labeled compounds.

F. Astatine

No stable nuclide of astatine exists. The longest-lived and the most useful nuclide is ^{211}At, which decays by EC (52%) and α-emission (42%) with 7.2 h halflife. Its α-particle energy is 5.87 MeV and its specific activity is 2.1×10^6 Ci/g. There are five other astatine nuclides with half-lives longer than 30 m: ^{206}At (EC and β^+; 31 m), ^{207}At (EC and α; 1.8 h), ^{208}At (EC; 1.6 h), ^{209}At (EC and α; 5.6 h), and ^{210}At (EC; 8.3 h).

Table 6
USEFUL CYCLOTRON NUCLIDES WITH MIDDLE AND RELATIVELY LONG LIVES

Nuclide	Half-life	Decay mode	Main γ-ray energy (MeV)	Production reaction	Reference no.	Commercial availability
^{7}Be	53 d	EC	0.478	^{7}Li(p,n)^{7}Be	69	Steady
^{22}Na	2.6 y	β$^+$(91%), EC	1.275	^{24}Mg(d,α) ^{22}Na	69	Steady
^{28}Mg	21 h	β$^-$	0.401	(See text)		Scarce
^{43}K	22 h	β$^-$	0.373	(See text)		Scarce
^{48}V	16 d	EC(50%), β$^+$	0.984	^{48}Ti(p,n)^{48}V	69	Occasional
^{48}Cr	22 h	EC	0.305	^{48}Ti(^{3}He,3n)^{48}Cr	70	No
^{52}Mn	5.6 d	EC(72%), β$^+$	1.434	^{52}Cr(d,2n) ^{52}Mn	69	Occasional
^{54}Mn	312 d	EC	0.835	^{56}Fe(d,α)^{54}Mn	69	Occasional
^{52}Fe	8.3 h	β$^+$(57%), EC	0.169	(See text)		Scarce
^{55}Co	17.6 h	β$^+$(77%), EC	0.477	^{56}Fe(p,2n)^{55}Co	71	No
^{57}Co	271 d	EC	0.122	^{58}Ni(p,2p)^{57}Co	69	Steady
^{57}Ni	36 h	EC(60%), β$^+$	1.378	^{56}Fe(^{3}He,2n)^{57}Ni	72	No
^{67}Cu	62 h	β$^-$	0.1845	Zn + d → ^{67}Cu	73	No
^{62}Zn	9.1 h	EC(93%), β$^+$	0.597	^{63}Cu(p,2n)^{62}Zn	74	No
^{67}Ga	78 h	EC	0.0933	(See text)		Steady
^{74}As	17.8 d	EC(37%), β$^+$, β$^-$	0.596	^{74}Ge(d,2n)^{74}As	69	Occasional
^{73}Se	7.2 h	β$^+$(65%), EC	0.361	(See text)		No
^{97}Ru	2.9 d	EC	0.216	Mo + α → ^{97}Ru	75	Scarce
^{101m}Rh	4.3 d	EC(93%), IT	0.307	^{103}Rh(p,3n)^{101}Pd → ^{101m}Rh	76	No
^{111}In	2.8 d	EC	0.171	(See text)		Steady
^{133m}Ba	39 h	IT	0.276	^{133}Cs(d,2n)^{133m}Ba	77	No
^{167}Tm	9.3 d	EC	0.208	^{165}Ho(α,2n)^{167}Tm	78	No
^{201}Tl	73 h	EC	Hg X-Ray	(See text)		Steady
^{203}Pb	52 h	EC	0.279	(See text)		Occasional
^{237}Pu	45 d	EC	Np X-Ray	(See text)		No

Astatine has been produced and its in vivo behavior studied from academic interest and for the possibility of being used as an internal radiation source in radiation therapy. If an astatine compound is selectively accumulated in the target position of the radiation therapy, the α-rays from astatine would serve effectively for the therapy.

^{211}At is produced efficiently by the ^{209}Bi (α,2n)^{211}At reaction. Its excitation function has been known for a long time.[49] Dry vaporization as well as distillation and solvent extraction are used for the separation of astatine.[50] Often iodine is added as non-nuclidic carrier, but in this case AtI is formed and the iodine serves as the carrier of AtI. Details about the production, separation and estimation of astatine are described in Reference 50. Some biomolecules, such as tyrosine, imidazole, and proteins, were labeled with ^{211}At, and their in vivo behavior was studied.[51,52]

II. MIDDLE AND RELATIVELY LONG-LIVED CYCLOTRON RADIONUCLIDES USED IN BIOLOGY AND MEDICINE

A. Useful Nuclides

The term ''middle and relatively long-lived'' is rather ambiguous, but here are surveyed useful cyclotron-made nuclides with half-lives longer than 6 h. Many of them are shown in Table 6. Generator nuclides, such as ^{68}Ge-^{68}Ga, ^{87}Y-^{87m}Sr, ^{72}Se-^{72}As and ^{178}W-^{178}Ta, as well as radiohalogens are excluded, because there are other chapters for them. Several nuclides in Table 6 are steadily available commercially. It is usually wiser for the users of them to buy rather than produce by themselves. Only the principle of production is given here for these commercial nuclides. Some nuclides of biological, medical, or environmental interest with half-lives less than 24 h are supplied only occasionally to limited parts of the world. They include ^{28}Mg, ^{43}K, and ^{52}Fe.

Different from the demand in diagnostic use, longer-lived γ-ray emitters are required or preferred in agricultural tracer use. For some elements, such a nuclide is not formed by reactor irradiation and should be produced by accelerator bombardment. ^{7}Be, ^{22}Na, ^{48}V, ^{54}Mn, ^{56}Ni, ^{67}Cu, ^{74}As, and ^{203}Pb belong to this category.

Some accelerator-made nuclides are also useful as radiation sources. As the positron source, ^{22}Na and ^{68}Ge-^{68}Ga are used. ^{57}Co is the most popular source in Mössbauer spectroscopy. Mössbauer spectroscopy with this source gives information about the chemical state of iron in various substances including biological samples.[53] When a nuclide emits two γ-rays successively, the angular correlation between them sometimes offers information about the microscopical environment of this nuclide.[54] ^{111}In emits γ-rays of 0.117 MeV and 0.247 MeV, and is a useful source for the angular correlation study.

B. ^{67}Ga, ^{111}In, and ^{201}Tl

These three nuclides are always available commercially in satisfactory purities and used for routine diagnosis. The reactions ^{68}Zn(p,2n)^{67}Ga, ^{67}Zn(d,2n)^{67}Ga, and ^{68}Zn(d,3n)^{67}Ga on enriched targets are preferred in routine commercial production of ^{67}Ga, though α-particle reactions on natural copper and zinc can also be used.[55] Protons and deuterons lose much smaller fractions of their energies than α-particles in a given thickness of a target substance, and thus gives a much higher yield in this production. The target substance can be electroplated on a metal pipe, through which cooling water is passed during the bombardment. Solvent extraction or ion exchange sometimes after coprecipitation with Fe(OH)$_3$ is used for the separation of the ^{67}Ga. Thus, almost 1 Ci of ^{67}Ga can be produced in one batch.

For the commercial production of ^{111}In, too, the ^{112}Cd(p,2n)^{111}In reaction on enriched ^{112}Cd(24% natural abundance) is preferred to the ^{109}Ag (α,2n)^{111}In reaction. Similar techniques are used in the bombardment and separation as in ^{67}Ga.[55]

^{201}Tl is now widely used for heart diagnosis, though its history in nuclear medicine is relatively short. Since there are several neutron-deficient radiothalliums with half-lives longer than several hours, ^{201}Tl for injection should be produced under well selected conditions. The ^{203}Tl(p,3n)^{201}Pb $\xrightarrow[9.4\ hr]{EC}$ ^{201}Tl is the best production reaction, the ^{201}Pb being separated from the target and left to decay into ^{201}Tl. There are, however, following reactions to give radiothalliums via radioleads: ^{198}Pb(2.4 h) → ^{198}Tl(5.3 h), ^{199}Pb(90 m) → ^{199}Tl(7.4 h), ^{200}Pb(21.5 h) → ^{200}Tl(26.1 h) and ^{202m}Pd(3.6 h, EC 10%) → ^{202}Tl (12 d). Excitation functions have been measured for the ^{203}Tl(p,3n)^{201}Pb reaction and for impurity formation reactions in order for the selection of optimum bombardment conditions.[56]

C. ^{73}Se

Selenium is regarded as an essential bioelement, and some seleno-organic compounds, such as selenomethionine, exist in animal body. Selenomethionine labeled with reactor-made ^{75}Se has been used for pancreas diagnosis in spite of its long life (120 d). ^{73}Se is a short lived positron emitter obtainable in no-carrier-added state. It is thus more suitable in nuclear medicine, though it decays into radioactive ^{73}As (EC, 80 d). Production of ^{73}Se as well as organic labeling with it have been studied. The reactions Ge + ^{3}He → ^{73}Se, Ge + α → ^{73}Se and ^{75}As(p,3n)^{73}Se can be used. The excitation functions for them were measured together with those for simultaneous formation of ^{72}Se (8.4 d) and ^{75}Se.[57] The ^{75}As(p,3n)^{73}Se reaction gives the highest yield and product purity. The thick target yields for these reactions are shown in Table 7.[57] As the germanium target, metallic germanium or GeO$_2$ are used. For arsenic target, some descriptions are given in Section I.D.3. From the targets, ^{73}Se can be separated either by volatillization under radio-frequency fusion or by some other wet processes.[57,58] With a minute amount (several micrograms) of carrier, ^{73}Se can be obtained as red selenium solution in an organic solvent.[57]

Table 7
THICK TARGET SATURATION ACTIVITY OF ^{73}Se, ^{72}Se AND ^{75}Se
(10^9 DPS/μA). (TARGET: ELEMENT OF NATURAL NUCLIDIC COMPOSITION)

Incident energy (MeV)		5	10	12.5	15	20	25	30	35	40	45	50
As + p	^{73}Se	0	0	0	0	0	0.1	3.5	11	19	25	29
	^{72}Se	0	0	0	0	0	0	0	0.01	0.56	2.9	6.4
	^{75}Se	1.1	6.9	10.5	12.3							
Ge + ^{3}He	^{73}Se	0	<0.01	0.01	0.02	0.11	0.20	0.30	0.35	0.50		
	^{72}Se	0	0	<0.01	0.01	0.03	0.06	0.14	0.24	0.33		
	^{75}Se						0.29			0.50		
Ge + α	^{73}Se	0	0	<0.01	0.03	0.15	0.22	0.26	0.28	0.37		
	^{72}Se	0	0	0	0	0	0.02	0.06	0.13	0.19		
	^{75}Se	0	0	0.01	0.05	0.22	0.37					

D. ^{52}Fe

The reactions ^{50}Cr(α,2n)^{52}Fe, ^{52}Cr(^{3}He,3n)^{52}Fe and ^{55}Mn(p,4n)^{52}Fe are used for the production of ^{52}Fe,[59-61] which is the only useful radio-iron for diagnostic use. The excitation functions for the chromium reactions are known.[61] Cross sections are low for all the three reactions, not exceeding 25 mb. Metallic chromium in pressed powder or electro-plated layer is used as the chromium target and bombarded with a maximum beam flux. In the α-particle bombardment, an enriched ^{52}Cr target (4.3% natural abundance) is usually used. The bombarded chromium target is dissolved in hydrochloric acid, and the ^{52}Fe after being oxidized to Fe^{+3} state is extracted into ether and back-extracted into water. In the manganese reaction, metallic manganese or MnO_2 is bombarded with protons of over 50 MeV energy. For the separation of the ^{52}Fe, the anion exchange from hydrochloric acid solution is also available. Care should be taken to minimize the contamination of nonradioactive iron from the target substance and reagents. With this contamination, ^{52}Fe when injected intravenously is often trapped in lung and liver.

E. ^{28}Mg and ^{43}K

These two nuclides are neutron excess and produced by proton emission reactions. Only ^{28}Mg can practically be used as radiotracer of magnesium, which is one of the most important elements in life sciences and metallurgy. Formerly ^{43}K was used for heart diagnosis. Now this use has been replaced by ^{201}Tl, but ^{43}K is still useful in the study on potassium fertilizers, etc.

The following reactions have been studied for the production of ^{28}Mg: ^{26}Mg(t,p)^{28}Mg, ^{26}Mg(α,2p)^{28}Mg, ^{27}Al(t,2p)^{28}Mg, ^{27}Al(α,3p) ^{28}Mg, and high energy proton on P, S, Cl, and Ar.[62-64] The triton reactions give higher yields than the α-particle reactions, but the acceleration of triton encounters various problems. The reactions on magnesium gives ^{28}Mg of limited specific activities. By any bombardment, ^{28}Mg cannot be obtained in a high yield, with simultaneous formation of much higher activity of ^{24}Na. The adsorption of magnesium on quartz powder can be used in the separation of ^{28}Mg from aluminum target.[62]

The ^{40}Ar(α,p)^{43}K reaction is used for the production of ^{43}K.[65] The excitation functions were measured for the ^{40}Ar(α,p)^{43}K and ^{40}Ar(α,pn)^{42}K reactions.[66] In order for ^{43}K to be used in diagnosis, the contamination of ^{42}K (12.5 h, energetic β$^-$,γ) should be minimized. Thus the incident particle energy should be made lower than 22 MeV with the sacrifice of the yield. An argon stream is used as the target. Two methods have been used for the trapping of the ^{43}K: (1) from a slow argon stream, ^{43}K is spontaneously trapped on the inside wall of the bombardment box, and (2) with a fast stream ^{43}K is carried out from the box, and trapped on a filter in the stream path. The adsorbed ^{43}K is then eluted with water. No high yield of ^{43}K can be expected. In practical production about 10 mCi is a maximum yield in one batch, even when a considerable contamination of ^{42}K is permissible. Better yields of ^{43}K can be obtained by the high energy deuteron reaction on titanium or high energy spallation reactions on some elements. However, the separation of ^{43}K in salt-free solution is very difficult from the targets.

F. ^{203}Pb and ^{237}Pu

^{203}Pb is useful for pollution study as well as in nuclear medicine. It is formed with a good yield by the ^{205}Tl(p,3n) ^{203}Pb reaction, often as the by-product in the ^{201}Tl production by the ^{203}Tl(p,3n)^{201}Pb → ^{201}Tl reaction.[56] (Nuclidic abundance of Tl: ^{203}Tl, 29.5%; ^{205}Tl, 70.5%)

^{237}Pu emits X-rays of neptunium (about 100 keV), and is the most suitable tracer nuclide of plutonium. It is produced either by the ^{235}U (α,2n) ^{237}Pu reaction on enriched ^{235}U or by the ^{238}U(^{3}He,4n)^{237}Pu reaction.[67] The yield is low for the former reaction and still poorer for the latter reaction, with simultaneous formation of much higher activities of fission products. Careful ion-exchange processes are necessary for the separation of the ^{237}Pu. It is difficult to obtain 1 mCi of ^{237}Pu by continuous bombardment for 1 day.

G. Other Nuclides

In Table 6 are also listed various other nuclides. Beside the production reactions in Table 6, high energy spallation reactions yield various nuclides including those which are not otherwise produced efficiently (e.g., ^{67}Cu and ^{52}Fe).[68] However, their chemical separation into a suitable state is usually difficult and tedious and their radionuclidic purity is often problematic.

Production methods of some of the nuclides in Table 6 were already established by 1960. They are given in Reference 69. For the other nuclides in Table 6 recent publications in well-circulating journals are cited there, so the reader can easily find out more detailed descriptions by consulting references in them. Often nuclear reactions other than in Table 6 can also be used, and in a given domestic nuclide production the most suitable reaction should be selected on the basis principally of the characteristics of the available cyclotron.

REFERENCES

1. **Nozaki, T., Iwamoto, M., and Ido, T.,** Yield of ^{18}F for various reactions from oxygen and neon, *Int. J. Appl. Radiat. Isotopes*, 25, 393, 1974.
2. **Fitschen, J., Beckmann, R., Holm, U., and Neuert, H.,** Yield and production of ^{18}F by ^{3}He irradiation of water, *Int. J. Appl. Radiat. Isotopes*, 28, 781, 1977.
3. **Ruth, T. J. and Wolf, A. P.,** Absolute cross section for the production of ^{18}F via the ^{18}O(p,n)^{18}F reaction, *Radiochim. Acta*, 26, 21, 1979.
4. **Friedlander, G., Kennedy, J. W., and Miller, J. M.,** *Nuclear and Radiochemistry*, 2nd ed., John Wiley & Sons, New York, 1964, 348.
5. **Backhausen, H., Stöcklin, G., and Weinreich, R.,** Formation of ^{18}F via its ^{18}Ne precursor, *Radiochem. Acta*, 29, 1, 1981.
6. **Clark, J. G. and Silvester, D. J.,** A cyclotron method for the production of fluorine-18, *Int. J. Appl. Radiat. Isotopes*, 17, 151, 1966.
7. **Helus, F., Wolber, G., Sahm, U., Abrams, D., and Maier-Borst, W.,** ^{18}F production methods, *J. Labeled Comp. Radiopharm.*, 16, 214, 1979.
8. **Irie, T., Fukushi, K., Ido, T., and Nozaki, T.,** Fluorine-18 fluorination in a carrier-free state by crown ether, *J. Labeled Comp. Radiopharm.*, 18, 9, 1981.
9. **Gatley, S. J. and Shaughnessy, W. J.,** Nucleophilic substitution with fluoride, *J. Labeled Comp. Radiopharm.*, 18, 24, 1981.

10. **Palmer, A. J., Clark, J. C., and Goulding, R. W.,** The preparation of fluorine-18 labelled radiopharmaceuticals, *Int. J. Appl. Radiat. Isotopes,* 28, 53, 1977.
11. **Tewson, T. J., Welch, M. J., and Reichle, M. E.,** [^{18}F]-labeled 3-deoxy-3-fluoro-D-glucose, *J. Nucl. Med.,* 19, 1339, 1978.
11a. **Tewson, T. J. and Welch, M. J.,** Preparation and preliminary biodistribution of no-carrier-added ^{18}F fluorethanol, *J. Nucl. Med.,* 21, 559, 1980.
12. **Casella, V., Ido, T., Wolf, A. P., et al.,** Anhydrous F-18 labeled elemental fluorine for radiopharmaceutical preparation, *J. Nucl. Med.,* 21, 750, 1980.
13. **Crouzel, C. and Comar, D.,** Production of carrier-free ^{18}F-hydrofluoric acid, *Int. J. Appl. Radiat. Isotopes,* 29, 407, 1978.
14. **Dahl, J. R., Lee, R., Schmall, B., and Bilger, R. E.,** A novel target for the preparation of anhydrous H^{18}F with no added carrier, *J. Labeled Comp. Radiopharm.,* 18, 34, 1981.
15. **Vine, E. N., Young, D., Vine, V. H., and Wolf, W.,** An improved synthesis of ^{18}F-5-fluorouracil, *Int. J. Appl. Radiat. Isotopes,* 30, 401, 1979.
16. **Ido, T., Van, C. N., Casella, V., et al.,** ^{18}F-labeled 2-deoxy-2-fluoro-D-glucose, 2-deoxy-2-fluoro-D-mannose, and ^{14}C-2-deoxy-2-fluoro-D-glucose, *J. Labeled Comp. Radiopharm.,* 14, 175, 1978.
17. **Bida, G. T., Ehrenkaufer, R. L., Wolf, A. P., et al.,** The effect of target gas purity on the chemical form of fluorine-18 during ^{18}F-F$_2$ production using the neon/fluorine target, *J. Nucl. Med.,* 21, 758, 1980.
18. **Wieland, B. M., Schlyer, D. J., Ruth, T. L., and Wolf, A. P.,** Deuteron beam penetration in a neon gas target for producing fluorine-18, *J. Labeled Comp. Radiopharm.,* 18, 27, 1981.
19. **Palmer, A. J.,** Recoil labelling of fluorine-18 labelled chlorofluoromethanes and tetrafluoromethane, *Int. J. Appl. Radiat. Isotopes,* 29, 4, 1978.
20. **Leowitz, E., Richard, R., and Baranowsky, J.,** ^{18}F recoil labeling, *Int. J. Appl. Radiat. Isotopes,* 23, 392, 1972.
21. **Lambrecht, R. M., Neirinchx, R., and Wolf, A. R.,** Novel anhydrous ^{18}F-fluorinating intermediates, *Int. J. Appl. Radiat. Isotopes,* 29, 175, 1978.
22. **Zatolokin, B. V., Konstantinov, I. O., and Krasnov, N. N.,** Thick target yields of ^{34m}Cl and ^{38}Cl produced by various charged particles on phosphorus, sulphur and chlorine targets, *Int. J. Appl. Radiat. Isotopes,* 27, 159, 1978.
23. **Lee, D. M. and Markowitz, S. S.,** ^{3}He activation analysis for S, Cl, K, and Ca, *J. Radioanal. Chem.,* 19, 159, 1974.
24. **Weinreich, R., Qaim, S. M., and Stöcklin, G.,** New excitation functions for the production of medically useful halogen radioisotopes, *J. Labeled Comp. Radiopharm,* 13, 233, 1977.
25. **Nozaki, T., Iwamoto, M., and Itoh, Y.,** Production of ^{77}Br by various nuclear reactions, *Int. J. Appl. Radiat. Isotopes,* 30, 79, 1979.
26. **Qaim, S. M., Stöcklin, G., and Weinreich, R.,** Excitation functions for the formation of neutron deficient isotopes of bromine and krypton via high-energy deuteron induced reactions on bromine, *Int. J. Appl. Radiat. Isotopes,* 28, 947, 1977.
27. **Janssen, A. G. M., Van den Bosch, R. L. P., De Goeij, J. J. M., and Theelen, H. M. J.,** The reactions ^{77}Se(p,n) and ^{78}Se(p,2n) as production routes for ^{77}Br, *Int. J. Appl. Radiat. Isotopes,* 31, 405, 1980.
28. **Brinkman, G. A. and Lindner, L.,** Excitation functions for the production of ^{76}Kr and ^{77}Kr, *Int. J. Appl. Radiat. Isotopes,* 30, 190, 1979.
29. **Paans, A. M. J., Weeleweerd, J., Vaalburg, W., et al.,** Excitation functions for the production of ^{75}Br, *Int. J. Appl. Radiat. Isotopes,* 31, 267, 1980.
30. **Weinreich, R., Qaim, S. M., and Stöcklin, G.,** Comparative studies on the production of the positron emitter bromine-75 and phosphorus-30, *J. Labeled Comp. Radiopharm.,* 18, 201, 1981.
31. **Nunn, A. D. and Waters, S. L.,** Target materials for the cyclotron production of carrier-free ^{77}Br, *Int. J. Appl. Radiat. Isotopes,* 26, 731, 1975.
32. **Weinreich, R. and Blessing, G.,** Production of ^{77}Br at a medical compact cyclotron, *J. Labeled Comp. Radiopharm.,* 16, 222, 1979.
33. **Madhusdhan, C. P., Treves, S., Wolf, A. P., and Lambrecht, R. M.,** Improvements in ^{77}Br production and radiochemical separation from enriched ^{78}Se, *J. Radioanal. Chem.,* 53, 299, 1979.
34. **Helus, F.,** Preparation of carrier-free bromine-77 for medical use, *Radiochem. Radioanal. Lett.,* 3, 45, 1970.
35. **Lundqvist, H., Malmborg, P., Longestrom, B., and Chingmal, S. N.,** Simple production of ^{77}Br$^-$ and ^{123}I$^-$ and their use in the labelling of [^{77}Br]BrUdR and [^{123}I]IUdR, *Int. J. Appl. Radiat. Isotopes,* 30, 39, 1979.
36. **Stöcklin, G.,** Bromine-77 and iodine-123 radiopharmaceuticals, *Int. J. Appl. Radiat. Isotopes,* 28, 113, 1977.
37. **Van den Bosch, R., De Goeij, J. J. M., Van der Heide, J. J., et al.,** A new approach to target chemistry for the iodine-123 production via the ^{124}Te(p,2n) reaction, *Int. J. Appl. Radiat. Isotopes,* 28, 255, 1977.

38. **Acerbi, E., Birattari, C., Castiglioni, M., et al.**, Production of ^{123}I for medical purpose at the Milan AVF cyclotron, *Int. J. Appl. Radiat. Isotopes*, 26, 741, 1975.
39. **Kondo, K., Lambrecht, R. M., and Wolf, A. P.**, Excitation functions for the ^{124}Te(p,2n)^{123}I and ^{124}Te(p,n)^{124}I reactions and the effect of target enrichment on radionuclidic purity, *Int. J. Appl. Radiat. Isotopes*, 28, 395, 1977.
40. **Kondo, K., Lambrecht, R. M., Norton, E. F., and Wolf, A. P.**, Improved target and radiochemistry for production of ^{123}I and ^{124}I, *Int. J. Appl. Radiat. Isotopes*, 28, 756, 1977.
41. **Syme, S. M., Wood, E., Blair, I. M., et al.**, Yield curves for cyclotron production of ^{123}I, ^{125}I and ^{121}I by ^{127}I(p,xn)Xe → (β) → I reaction, *Int. J. Appl. Radiat. Isotopes*, 29, 29, 1978.
42. **Weinreich, R., Schult, O., and Stöcklin, G.**, Production of ^{123}I via the ^{127}I(d,6n)^{123}Xe(β$^+$, EC)^{123}I process, *Int. J. Appl. Radiat. Isotopes*, 25, 535, 1974.
43. **Lambrecht, R. M., Wolf, A. P., Helus, F., et al.**, High energy alpha reactions for the production of the ^{123}Cs → ^{123}Xe → ^{123}I generator and ^{123}Xe and ^{127}Cs for radiopharmaceutical application, *Int. J. Appl. Radiat. Isotopes*, 27, 675, 1976.
44. **Peek, N. F. and Hegedus, F.**, The production of xenon isotopes with protons of energies from 320 to 590 MeV, *Int. J. Appl. Radiat. Isotopes*, 30, 631, 1979.
45. **Adilbish, M., Chumin, V. G., Kahlkin, V. A., et al.**, ^{123}I production from radioxenon formed in spallation reactions by 660 MeV protons for medical research, *Int. J. Appl. Radiat. Isotopes*, 31, 163, 1980.
46. **Cuninghame, J. G., Morris, B., Nichols, A. L., and Taylor, N. K.**, Large scale production of ^{123}I from a flowing liquid target using the (p,5n) reaction, *Int. J. Appl. Radiat. Isotopes*, 27, 597, 1976.
47. **Shimmel, A., Kaspersen, F. M., and Lindner, L.**, Cyclotron production of ^{123}Xe (→ ^{123}I) and ^{127}Xe using CH$_2$I$_2$ as target material, *Int. J. Appl. Radiat. Isotopes*, 30, 63, 1979.
48. **Godart, J., Barat, J. L., Menthe, A., and Briere, J.**, In beam collection of ^{123}Xe for carrier-free ^{123}I production, *Int. J. Appl. Radiat. Isotopes*, 28, 967, 1977.
49. **Ramler, W. J., Wing, J., Henderson, D. J., and Huizenga, J. R.**, Excitation functions of bismuth and lead, *Phys. Rev.*, 114, 154, 1959.
50. **Downs, A. J. and Adams, C. J.**, Preparation, separation and estimation of astatine, in *Comprehensive Inorganic Chemistry*, Vol. 2, Blailar, J. C., Emeleus, H., Nyholm, R., and Trotman-Dickenson, A. F., Eds., Pergamon Press, Oxford, 1973, 1582.
51. **Lisser, G. W. M., Diemer, E. L., and Kaspersen, F. M.**, The preparation and stability of astatotyrosine and astatoiodotyrosine, *Int. J. Appl. Radiat. Isotopes*, 30, 749, 1979.
52. **Vaughan, A. T. M.**, The labelling of proteins with ^{211}At using an acylation reaction, *Int. J. Appl. Radiat. Isotopes*, 30, 576, 1979.
53. **Johnson, E. C.**, Mössbauer spectroscopy, in *Topics of Applied Physics*, Vol. 5, Gonser, U., Ed., Springer-Verlag, Berlin, 1975, chap. 4.
54. **Adloff, J. P.**, Application to chemistry of electric quadrupole perturbation of γ-γ angular correlation, *Radiochim. Acta*, 25, 57, 1978.
55. **Thakur, M. L.**, Gallium-67 and indium-111 radiopharmaceuticals, *Int. J. Appl. Radiat. Isotopes*, 28, 183, 1977.
56. **Lugunas-Solar, M. C., Jungerman, J. A., Peek, N. F., and Theus, R. M.**, Thallium-201 yields and excitation functions for the lead radioactivities produced by irradiation of natural thallium with 15—60 MeV protons, *Int. J. Appl. Radiat. Isotopes*, 29, 159, 1978.
57. **Nozaki, T., Itoh, Y., and Ogawa, K.**, Yield of ^{73}Se for various reactions and its chemical processing, *Int. J. Appl. Radiat. Isotopes*, 30, 595, 1979.
58. **Guillaume, M., Lambrecht, R. M., Christiaens, L., et al.**, Carrier-free selenium-73: cyclotron production and seleno-radiopharmaceuticals synthesis, *J. Labeled Comp. Radiopharm.*, 16, 126, 1979.
59. **Thakur, M. L., Nunn, A. D., and Waters, S. L.**, Iron-52: improving its recovery from cyclotron targets, *Int. J. Appl. Radiat. Isotopes*, 22, 481, 1971.
60. **Akiha, M., Aburai, T., Nozaki, T., and Murakami, Y.**, Yield of ^{52}Fe for the reaction of ^{3}He and α on chromium, *Radiochim. Acta*, 18, 108, 1972.
61. **Saha, G. B. and Farrer, P. A.**, Production of ^{52}Fe by the ^{55}Mn(p,4n)^{52}Fe reaction for medical use, *Int. J. Appl. Radiat. Isotopes*, 22, 495, 1971.
62. **Nozaki, T., Furukawa, M., Kume, S., and Seki, R.**, Production of ^{28}Mg by triton and α-particle indiced reactions, *Int. J. Appl. Radiat. Isotopes*, 26, 17, 1975.
63. **Probst, H. J., Qaim, S. M., and Weinreich, R.**, Excitation functions of high-energy α-particle induced nuclear reactions on aluminium and magnesium, *Int. J. Appl. Radiat. Isotopes*, 27, 431, 1976.
64. **Lungqvist, H. and Malmborg, P.**, Production of carrier-free ^{28}Mg and ^{24}Na by 50—180 MeV protons on Si, P, S, Cl, Ar, and K, *Int. J. Appl. Radiat. Isotopes*, 30, 33, 1979.
65. **Clark, J. G., Thakur, M. L., and Watson, I. A.**, The production of potassium-43 for medical use, *Int. J. Appl. Radiat. Isotopes*, 23, 329, 1972.
66. **Tanaka, S., Furukawa, M., Mikumo, T., et al.**, Reactions of argon-40 with alpha-particles, *J. Phys. Soc. Japan*, 15, 952, 1960.

67. **Hata, K., Baba, H., Umezawa, H., et al.,** Preparation of a gamma-ray emitting plutonium isotope ^{237}Pu, *Int. J. Appl. Radiat. Isotopes,* 27, 713, 1976.
68. **Grant, P. M., O'Brien, H. A., Bayhurst, B. P., Jr., et al.,** Spallation yields of Fe-52, Cu-67 and Tl-201 from reactions of 800 MeV protons with Ni, As, Pb and Bi targets, *J. Labeled Comp. Radiopharm.,* 16, 213, 1979.
69. **Gruverman, I. J. and Kruger, P.,** Cyclotron-produced carrier-free radioisotopes, *Int. J. Appl. Radiat. Isotopes,* 5, 21, 1959.
70. **Weinreich, R., Probst, H. J., and Qaim, S. M.,** Production of ^{48}Cr for applications in life sciences, *Int. J. Appl. Radiat. Isotopes,* 31, 223, 1980.
71. **Lugunas-Solar, M. C. and Jungerman, J. A.,** Cyclotron production of carrier-free ^{55}Co, *Int. J. Appl. Radiat. Isotopes,* 30, 25, 1979.
72. **Neirinckx, R. D.,** Cyclotron production of Ni-57 and Co-55, *Int. J. Appl. Radiat. Isotopes,* 28, 561, 1977.
73. **Neirinckx, R. D.,** Simultaneous production of ^{67}Cu and ^{64}Cu and ^{67}Ga and labelling of bleomycin with ^{67}Cu and ^{64}Cu, *Int. J. Appl. Radiat. Isotopes,* 28, 208, 1977.
74. **Suzuki, K. and Iwata, R.,** Simultaneous production of ^{123}I, ^{62}Zn and ^{13}NH$_3$, *Int. J. Appl. Radiat. Isotopes,* 28, 663, 1977.
75. **Silvester, D. J., Helus, F., and Maier-Borst, W.,** Ruthenium-97: half-life, yield measurements and isolation from molybdenum targets, *J. Labeled Comp. Radiopharm.,* 16, 226, 1979.
76. **Scholz, K. L. and Sodd, V. J.,** Cyclotron production of rhodium-101m through its precursor palladium-101, *Int. J. Appl. Radiat. Isotopes,* 28, 207, 1977.
77. **Neirinckx, R. D.,** Production of ^{133}Ba for medical purpose, *Int. J. Appl. Radiat. Isotopes,* 28, 323, 1977.
78. **Homma, Y., Sugitani, Y., Matsui, Y., et al.,** Cyclotron production of ^{167}Tm from natural erbium and natural holmium, *Int. J. Appl. Radiat. Isotopes,* 31, 505, 1980.

Chapter 4

THE SPECIAL POSITION OF ^{99m}Tc IN NUCLEAR MEDICINE

R. E. Boyd

TABLE OF CONTENTS

I.	Introduction 126	
	A.	Physical Properties of ^{99m}Tc 126
	B.	The Suitability of ^{99m}Tc for Organ Imaging........................... 126
	C.	Supply Considerations for ^{99m}Tc; The Generator Concept 127
	D.	The Kinetics of Growth and Decay of the ^{99}Mo:^{99m}Tc System.......... 127
II.	The Production of Molybdenum-99 129	
	A.	Neutron Activation of Molybdenum 129
	B.	Molybdenum-99 from the Fission of Uranium 130
	C.	Choosing the Method of Producing ^{99}Mo 132
III.	Methods of Separating ^{99m}Tc from ^{99}Mo 133	
	A.	Chromatographic Generators of ^{99m}Tc 133
	B.	Concentration Profile 133
	C.	Elution Efficiency 133
	D.	Radionuclide Purity.. 135
	E.	Radiochemical Purity 137
	F.	Chemical Purity 137
	G.	Biological Purity... 138
IV.	The Sublimation Generator.. 139	
	A.	Radionuclidic Purity of Sublimed ^{99m}Tc 140
	B.	Other Purity Aspects Associated with Sublimed ^{99m}Tc................. 142
V.	The Solvent Extraction Generator 142	
	A.	The Efficiency of the Solvent Extraction Generator 144
	B.	Radionuclidic Purity of Solvent-Extracted ^{99m}Tc....................... 144
	C.	Chemical Purity of Solvent-Extracted ^{99m}Tc 145
VI.	Comparison of ^{99m}Tc-Generator Forms....................................... 145	
VII.	Radiopharmaceuticals of ^{99m}Tc 147	
	A.	Technetium Chemistry.. 147
	B.	Technetium Radiopharmaceuticals 148
References.. 149		

I. INTRODUCTION

Technetium, number 43 in the periodic table, was the first element to be made artificially. Its discovery in 1937 was followed by a lengthy period in which it remained somewhat of a scientific curiosity. In recent times, however, it has been at the center of a revolution in the techniques used in medical diagnosis and has subsequently become an essential commodity in all technologically developed societies.

Initially, the element was identified in a piece of molybdenum which had been bombarded with protons in a cyclotron.[1,2] Subsequently it was produced in a nuclear reactor either by thermal neutron activation of molybdenum or by the fission of uranium. In all, twenty-one isotopes of technetium have been identified; all are radioactive with half-lives ranging from less than one second (^{110}Tc) to over four million years (^{98}Tc).[3]

Although the element is claimed to be a remarkable corrosion inhibitor for steel, its cost and, more importantly, its radioactivity have prevented technetium being put to any useful industrial purpose. However, when the selection criteria to be applied to radionuclides for in vivo use were being considered, it was discovered that one radionuclide of technetium, namely ^{99m}Tc, possessed a comparatively rare combination of properties which made it ideal for the organ imaging techniques that were being developed in nuclear medicine.

A direct result of the application of ^{99m}Tc, in conjunction with instruments capable of mapping its distribution within the human body, is the universal acceptance of nuclear techniques in medical diagnosis. Technetium-99m now occupies a position of preeminence as the most frequently used radionuclide in routine clinical practice.

A. Physical Properties of ^{99m}Tc

Technetium-99m decays by isomeric transition according to the decay scheme shown in Figure 1. The principal gamma-emission associated with the transition to the ground state of ^{99}Tc is γ_2, 0.1404 MeV photons which occurs at a frequency of 0.89 per disintegration. Apart from a very low frequency, direct beta decay to ruthenium-99, ^{99}Ru (see Figure 2) particulate radiation associated with the decay of ^{99m}Tc is limited to internal conversion and auger electrons.

The half-life of ^{99m}Tc is 6.02 hr.

B. The Suitability of ^{99m}Tc for Organ Imaging

In techniques where a radioactive substance is internally administered, the patient is necessarily exposed to potentially harmful radiation. The extent of the radiation exposure depends upon:

- How much of the radionuclide is administered
- The type and quality of the radiation emitted by the radionuclide
- How long the radionuclide remains within the patient

The justification for radiation exposure is a matter of concern to the attendant physician who must decide whether submitting the patient to a test entails greater risk than not performing the test. Because of ^{99m}Tc's unique physical properties, its use incurs only a small radiation dose to the patient.

The effectiveness of imaging instruments used to plot the distribution of the radionuclide within the patient's body is also influenced by the radioactive properties of the substance administered. Since most imaging instruments use only a thin, thallium activated, sodium iodide crystal as the detector, detection efficiency tends to fall away with increasing photon energy.

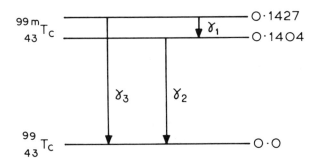

FIGURE 1. Isomeric level decay of ^{99m}Tc.

Determination of the origin in space of a photon detected by the NaI(Tl) crystal requires a lead collimator placed in front of the detector. The design of this device is governed by the energy of the emitted photons; efficient collimation is lost as the photon energy is increased.

The radiation emitted by ^{99m}Tc may be collimated and detected with maximum efficiency. (NB — The radiation emitted must have sufficient penetration that it can be detected, even when originating from deep-seated tissue; ^{99m}Tc adequately satisfies this requirement.)

The very favorable properties of ^{99m}Tc enable the physician to administer large quantities of the radionuclide to improve counting statistics, spatial resolution and hence, the general quality of the images, without incurring excessive radiation dose to the patient.

C. Supply Considerations for ^{99m}Tc; The Generator Concept

An important property of ^{99m}Tc, essential to reducing radiation exposure, is its half-life of 6 hr. Normally it would be expected that such a short-lived substance would present considerable practical problems of supply and distribution. Fortunately, however, ^{99m}Tc possesses a further attribute in that it is formed by the decay of molybdenum-99 (^{99}Mo; $T_{0.5}$ = 66 hr). The opportunity to provide ^{99m}Tc in the form of its longer-lived parent enables this radionuclide to be used at great distances from the site of production.

Considerable work has gone into the development of practical devices which permit ^{99m}Tc to be separated from ^{99}Mo. However, because of the nature of the parent:daughter radionuclide relationship that exists between ^{99}Mo and ^{99m}Tc, a single separation of ^{99m}Tc may, at later times, be followed by a second, third, and subsequent separations from the same quantity of ^{99}Mo; the exploitation of this phenomenon has led to the creation of the ^{99}Mo:^{99m}Tc generator.

Because these generators have played such an important role in making available the benefits inherent in ^{99m}Tc, this review will concentrate on them and their development and will describe the various options together with the relative advantages and disadvantages.

D. The Kinetics of Growth and Decay of the ^{99}Mo:^{99m}Tc System

In a system where one radioactive species decays to form another, certain mathematical expressions define the quantities (atoms or units of radioactivity) of each species present at any time.

The behavior of the parent radioactivity obeys the simple radioactive decay law,

$$\frac{dN_1}{dt} = -\lambda_1 N_1 \qquad (1)$$

or

$$(N_1)_t = (N_1)_0 \, e^{-\lambda_1 t} \qquad (2)$$

where $(N_1)_t$ = No. of parent atoms at time t, and λ_1 = Decay constant of the parent.
The number of atoms of the daughter radioactivity at time t is given by

$$(N_2)_t = \frac{\lambda_1}{\lambda_2 - \lambda_1} \cdot (N_1)_0 \cdot (e^{-\lambda_1 t} - e^{-\lambda_2 t}) + (N_2)_0 \cdot e^{-\lambda_2 t} \quad (3)$$

where $(N_2)_t$ = No. of daughter atoms at time t, and λ_2 = Decay constant of the daughter.
Converting Equations 2 and 3 from numbers of atoms to units of radioactivity we obtain

$$(A_1)_t = (A_1)_0 \, e^{-\lambda_1 t} \quad (4)$$

where $(A_1)_t$ = Radioactivity of parent at time t, and

$$(A_2)_t = \frac{\lambda_2}{\lambda_2 - \lambda_1} \cdot (A_1)_0 \cdot (e^{-\lambda_1 t} - e^{-\lambda_2 t}) + (A_2)_0 \cdot e^{-\lambda_2 t} \quad (5)$$

where $(A_2)_t$ = Radioactivity of daughter at time t.

The effect and solution of Equations 4 and 5 depends on the relative values of the decay constants λ_1 and λ_2 (and hence the half-lives, since $\lambda = 0.693/T_{0.5}$) of the parent and daughter species.

If λ_1 is very much smaller than λ_2 (and $T_{0.5}(1) \gg T_{0.5}(2)$), the system enters into secular equilibrium where the daughter activity eventually grows to equal the parent activity.

If the ratio $\lambda_1:\lambda_2$ is larger, say 0.01-1, the parent-daughter system enters into a state of transient equilibrium in which

1. The daughter activity will achieve a value greater than that of the parent (assuming a decay scheme with no branching);
2. The daughter activity will reach a maximum value after which it will decline; and
3. An equilibrium between the activities of the parent and daughter species will be achieved where the rate of decay of the daughter is equal to its rate of formation from the parent. The daughter will appear to decay according to the half-life of the parent.

Finally, if λ_1 is greater than λ_2, then no equilibrium is achieved and the parent species decays rapidly to form the daughter.

Figure 2 shows the complex decay scheme for the ^{99}Mo:^{99m}Tc:^{99}Tc:^{99}Ru system with respective half-lives and branching factors. For ^{99}Mo and ^{99m}Tc the values of λ_1 and λ_2 are such that a state of transient equilibrium will be achieved after the ^{99m}Tc activity has gone through a maximum value. However, the ^{99m}Tc activity will never exceed the ^{99}Mo activity because only 87.5% of the disintegrations of ^{99}Mo result in ^{99m}Tc. If the appropriate values for λ_1 and λ_2 are substituted in Equation 5 and a correction applied for the branching decay, the growth of ^{99m}Tc from the decay of ^{99}Mo can be calculated from

$$(A_2)_t = 0.963 \, (A_1)_0 \, (e^{-0.0105t} - e^{-0.1151t}) + (A_2)_0 \cdot e^{-0.1151t} \quad (6)$$

Putting $(A_2)_0 = 0$ at $t = 0$ and differentiating, the time at which the maximum daughter activity occurs can be obtained;

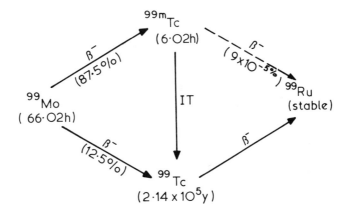

FIGURE 2. Decay sequences of ^{99}Mo to ^{99}Ru.

$$\frac{dA_2}{dt} = -0.963\,(A_1)_0\,[0.0105\,e^{-0.0105t} - 0.1151\,e^{-0.1151t}] \quad (7)$$

At $t = t_{max}$, $\frac{dA_2}{dt} = 0$, and hence

$$t_{max} = \frac{\ln(0.0105/0.1151)}{0.0105 - 0.1151}$$

$$= 22.89 \text{ hours.}$$

The decay-growth curve for the ^{99}Mo:^{99m}Tc system is shown in Figure 3.

The radioactivity attributable to ^{99}Tc is insignificant because of its very long half-life of 2.14×10^5 years; for example, 1 curie of ^{99}Mo will produce only 4×10^{-8} curies of ^{99}Tc.

II. THE PRODUCTION OF MOLYBDENUM-99

There are several methods of separating ^{99m}Tc from parent ^{99}Mo and in most cases the choice of method is dictated by the mode of preparation and hence quality of the ^{99}Mo.

Molybdenum-99 can be produced in two essentially different modes; one by the neutron activation of molybdenum, the other by uranium fission. These are described in practical terms.

A. Neutron Activation of Molybdenum

Molybdenum-99 may be produced by the irradiation of molybdenum with neutrons according to the reaction:

$$^{98}\text{Mo}\,(n,\gamma)\,^{99}\text{Mo}$$

The thermal neutron activation cross-section for this reaction is 0.14 barns; however, the effective (Westcott) cross-section can be significantly higher in a typical irradiation situation where there is a significant epithermal neutron flux giving rise to resonance neutron capture. Specific activities of the order of 1 Ci ^{99}Mo/g Mo are achievable from the irradiation of natural molybdenum in a high flux reactor. The use of enriched ^{98}Mo target material produces an increase in the specific activity which is in proportion to the enrichment factor and the change in effective cross-section due to different target geometry; under these circumstances the specific activity can be increased by a factor of approximately 8.

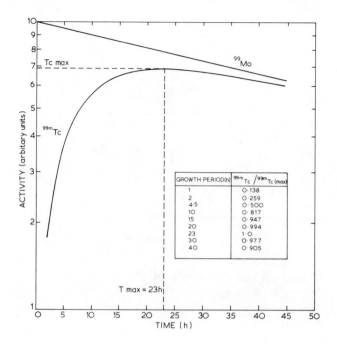

FIGURE 3. Decay-growth ^{99}Mo:^{99m}Tc system.

During the irradiation of molybdenum a number of concurrent nuclear reactions occur; some of these give rise to radionuclidic trace impurities which may be transferred to the ^{99m}Tc during a separation process; certain impurities commonly associated with molybdenum can be troublesome and must be minimized by careful selection of target quality.

The most frequently used target materials are molybdenum trioxide and molybdenum metal; both are suitable for use in high flux irradiations. Other molybdenum compounds (including some complexes for the enrichment of the ^{99}Mo by the Szillard-Chalmers reaction) have been employed in lower neutron fluxes.[4]

Molybdenum metal is not easily dissolved in mineral acids but as Ganiev et al.[5] have shown, the reaction between molybdenum and hydrogen peroxide solution produces a bright yellow solution (molybdenum peroxide) which gradually turns dark blue (molybdenum blue). Somerville[6] has indicated that this process is used commercially to dissolve irradiated molybdenum metal of very high purity. After reaction with hydrogen peroxide the Mo-species is converted to molybdate by the addition of sodium hydroxide.

Chemical processing of the irradiated molybdenum trioxide is limited to dissolution in sodium, potassium, or ammonium hydroxide, followed by either acidification to pH 1.5— 3 where the ^{99}Mo is destined for adsorption onto an alumina column, or adjustment of the excess alkalinity (range 0.5 - 5 M alkali) to produce the feedstock for the solvent extraction process. Alkaline solutions of radiomolybdate are, however, somewhat unstable and undergo radiation induced reduction, often precipitating a dark colored substance. Lewis and Tenorio[7] claim that this undesirable effect can be overcome by the addition of sodium hypochlorite to the ^{99}Mo solution.

When destined for use in a sublimation generator, the irradiated molybdenum trioxide requires no chemical processing after irradiation.

B. Molybdenum-99 from the Fission of Uranium

Molybdenum-99 is also a product of the fission of uranium according to

$$^{235}U\ (n,f)\ ^{99}Mo$$

The fission yield for this reaction is 6.1% and the irradiation of 1 g ^{235}U for 7 days in a neutron flux of 7×10^{13} n/cm^2/s produces approximately 142 curies ^{99}Mo.

The essential difference between ^{99}Mo produced by fission and that from the (n,γ) reaction is that its specific activity is very high. It is not, however, carrier free because other molybdenum isotopes are also formed by the fission of uranium, e.g.,

$$^{235}U(n,f) \cdots {}^{97}Zr \xrightarrow{94.6\%} {}^{97}Nb \text{ (60 sec)}$$
$$(16.8h) \xrightarrow{5.4\%} {}^{97}Nb (72.1m) \longrightarrow {}^{97}Mo \text{-fission yield} - 5.9\% \text{ (stable)}$$

$$^{235}U(n,f) \cdots {}^{98}Nb (51m) \longrightarrow {}^{98}Mo \text{-fission yield} - 5.9\% \text{ (stable)}$$

$$^{235}U(n,f) \cdots {}^{100}Mo \text{ - fission yield } - 6.3\% \text{ (stable)}$$

The net result of these reactions is a reduction in the specific activity of ^{99}Mo by an order of magnitude. A period of postirradiation decay produces a further reduction in specific activity.

In the choice of target material considerable attention must be paid to the effects of nuclear heating. It is essential that in the design of the target, provision is made for dissipating heat; insufficient heat removal, could cause the temperature of the target to exceed 1500°C and the risk that the target capsule will rupture and that some of the ^{99}Mo will sublime out of the target material, condense on the internal surfaces of the capsule, and be lost to subsequent extraction processes.

The target form can be varied by use of uranium enriched in ^{235}U; the presence of ^{238}U in the target material involves a risk of contaminating the ^{99}Mo product with highly toxic plutonium (^{239}Pu) formed by the reactions

$$^{238}U(n,\gamma) \, {}^{239}U \xrightarrow{\beta^-} {}^{239}Np \xrightarrow{\beta^-} {}^{239}Pu$$

The use of a target enriched in ^{235}U reduces or eliminates this risk.

Special precautions are necessary, however, with highly enriched targets to ensure adequate dissipation of the nuclear heat which may be more localized and hence very intense. This can be accomplished safely either by forming an alloy of the uranium with a much larger mass of aluminum or by dispersing the uranium over a large surface area as, say, an electroplated layer.

For uranium of low ^{235}U enrichment a suitable target is the dioxide. This material, after compressing and sintering into ceramic form, is both stable and a fair conductor of heat when irradiated.

A chemical process to extract curie quantities of ^{99}Mo from irradiated uranium was first described by the Brookhaven group.[8] In this process the target (93% enriched ^{235}U alloyed with Al) was dissolved in 6 M nitric acid catalyzed by mercuric nitrate. Then, after the addition of tellurium carrier, the solution was passed through an alumina column which selectively absorbed the ^{99}Mo and the radiotellurium fission products. The uranium and the unabsorbed fission products were removed from the alumina by washing. The ^{99}Mo was then recovered from the column by elution with $1M$ NH_4OH. An efficiency of approximately 70%, for a product purity of 99.99% ^{99}Mo, was claimed for this process. Richards[9] subsequently added a final purification step whereby the ^{99}Mo was readsorbed onto a strong anion exchange resin, washed to remove the trace radio-impurities and then eluted with $1.2M$ HCl.

The process used by the Australian Atomic Energy Commission is essentially the same as that used at Brookhaven except that the target used is uranium dioxide (2% enriched in ^{235}U) and the ^{99}Mo solution is subjected to a different purification sequence. In this case, the $1M$ NH_4OH eluate from the alumina column is passed through a 0.22 μm membrane filter to remove entrained particles of ^{132}Te-bearing alumina. The filtrate is then acidified with nitric acid, evaporated to dryness, and baked to a dull red heat. This treatment sublimes out the radioiodines and ruthenium-103. The neutralization of the ammoniacal eluate with nitric acid produces considerable quantities of ammonium nitrate. During the evaporation and baking processes the ammonium nitrate initially forms a voluminous residue. However, as the temperature is raised, this residue sublimes and decomposes according to

$$NH_4NO_3 \longrightarrow 2H_2O + N_2O$$

The resulting residue, which contains the ^{99}Mo, is essentially free of unnecessary chemical additives. The ^{99}Mo is finally dissolved by warming in $1M$ HNO_3.

The Oak Ridge group has developed an alternative process[10] for the extraction of ^{99}Mo from fissioned uranium. This method exploits the extractability of molybdenum from acid solution by di-(2-ethylhexyl) phosphoric acid dissolved in an inert organic diluent. A number of other methods for extracting fission product ^{99}Mo have been described.[11-14]

Silver-coated carbon granules have been shown to be most effective in separating ^{99}Mo from acid solutions of irradiated uranium.[15] In this process, the uranium dioxide is dissolved in $2N$ H_2SO_4 with added H_2O_2; after dissolution H_2SO_3 is added to destroy the residual H_2O_2. The solution is then added to a column of silver-coated charcoal which selectively extracts the ^{99}Mo from solution. After the column is washed with a small volume of water, the ^{99}Mo is eluted with $0.2\ M$ NaOH. Gamma spectrometric analysis has shown that the radionuclidic purity of the ^{99}Mo is greater than 99%, but that contamination by ^{132}Te-^{132}I and ^{103}Ru is also evident. The impurity levels, though small, would demand that the ^{99}Mo made by this process be further purified before being used for generator manufacture.

Silver-coated alumina may also be used to advantage in the extraction of ^{99}Mo from solutions of irradiated uranium.[16] Even in the presence of a high concentration of uranium (> 100 g L^{-1}), ^{99}Mo is quantitatively extracted from a solution of $1M$ nitric acid by silver-coated alumina. Desorption of molybdenum occurs when the silver-coated alumina column is eluted with ammonia; the efficiency of the recovery step being 75 to 80%. The purity of the ^{99}Mo eluted from silver-coated alumina is very high particularly as regards the ^{132}Te level.

C. Choosing the Method of Producing ^{99}Mo

Molybdenum-99 produced by the neutron activation of molybdenum requires minimum postirradiation processing and produces little radioactive waste. Although the activation cross-section is small, large quantities of molybdenum metal or molybdenum trioxide may be safely irradiated to give a high yield of ^{99}Mo. Neutron activated ^{99}Mo is always of low specific activity (< 10 Ci g^{-1} Mo).

Molybdenum-99 production from the fission of ^{235}U demands elaborate processing facilities of high capital cost; extreme precautions must be taken to avoid contaminating the product with toxic fission products and transuranics. Despite the high fission cross-section of ^{235}U and high fission yield of ^{99}Mo, the overall yield has to be restricted because of practical considerations, e.g., the ability to dissipate nuclear heat or to dispose of highly radioactive waste products. The specific activity of fission-produced ^{99}Mo is high ($> 10^4$ Ci g^{-1} Mo).

The criteria for choosing the method of ^{99}Mo-production to adopt must include economics, resources, and mode of utilization. The practical difficulties associated with the production of fission ^{99}Mo are reflected in the required capital investment and routine running costs; even in large-scale manufacture the cost of producing 1 Ci fission ^{99}Mo may be upwards of four times the cost of 1 Ci (n,γ) ^{99}Mo.

III. METHODS OF SEPARATING ^{99m}Tc FROM ^{99}Mo

Chromatography, sublimation, and solvent extraction are the three most common methods used to separate ^{99m}Tc from ^{99}Mo. Since the kinetics of the system provides for the regrowth of ^{99m}Tc after each separation cycle, the practical devices employed must be capable of repetitive operation for maximum effect. These devices, called generators, are of varied design. It is appropriate that each kind of generator be examined in detail to achieve a full understanding of the relative advantages and disadvantages.

A. Chromatographic Generators of ^{99m}Tc

The first ^{99m}Tc generator, developed at Brookhaven National Laboratory,[17] relied upon a chromatographic separation of ^{99m}Tc from ^{99}Mo; this method is still the most common. The technique is based on the relative differences in the distribution coefficients of aluminum oxide for the anions, molybdate, and pertechnetate. The passage of physiological saline through an alumina bed containing absorbed molybdate/pertechnetate will result in the elution of the pertechnetate component.

A critical assessment of the performance of the chromatographic generator includes a study of its elution concentration profile,[18] ^{99m}Tc elution efficiency,[19] the extent of radionuclidic contamination of the eluate, and finally the contamination of the eluate with the material of the bed itself.

B. Concentration Profile

A study of the effect of alumina bed size on the concentration profile of eluted ^{99m}Tc yielded the results shown in Figure 4. The elution profiles have gaussian shapes with decreasing peak heights and peak widths that increase with the size of the alumina bed. These findings emphasize the desirability of restricting bed size to a minimum; this in turn, influences the mode of production chosen for the ^{99}Mo. The low absorptive capacity of alumina for the molybdate ion dictates the use of a large bed when low specific activity ^{99}Mo is employed; conversely with carrier-free fission ^{99}Mo the bed size may be minimized.

C. Elution Efficiency

The elution efficiency of a generator can be defined simply as the proportion of the ^{99m}Tc present in the system that is separated during the elution process, usually expressed as a percentage. In practice the amount of ^{99m}Tc separated is frequently less than the amount predicted by theory (Equation 5) and much work has been performed to establish the factors that influence elution efficiency.

Vesely and Cifka[20] showed that the pertechnetate species could be reduced by ionizing radiation and that the lower valency states of Tc were strongly absorbed by the alumina. Strong circumstantial evidence exists that the generator is subject to attack by free radicals.

It has been well established that the species H_2, H_2O_2, H·, ·OH, e_{aq}^-, and H_3O^+ are formed in water exposed to ionizing radiation.[21] The free radical species formed are highly reactive and react with many solutes at rates which are close to diffusion controlled. In general, the most rapid reactions involving the reducing radical species, H· and e_{aq}^-, are with solutes

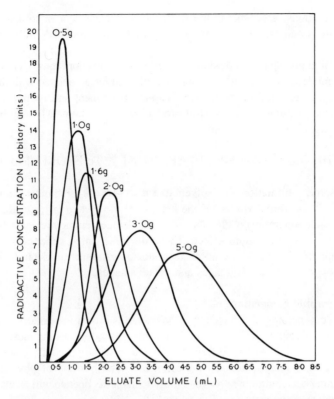

FIGURE 4. Elution profile vs. alumina bed weight. (Area under each curve normalized to 100 mCi ^{99m}Tc.)

which are readily reduced; whereas the most rapid reactions involving the oxidizing radical species, ·OH, are with solutes which are readily oxidized. Hydrogen gas usually does not participate in reactions with solutes but hydrogen peroxide reacts as an oxidizing agent or, under some circumstances, as a reducing agent.

Therefore, when an aqueous solution is irradiated, among the multitude of simultaneous reactions that occur are both oxidizing and reducing reactions. Whether the balance of reactions leads to net reduction, net oxidation, or a mixture of oxidized and reduced products depends very much on the nature of the solute.

The situation in a technetium generator is more complicated than in an aqueous solution. Matthews[22] speculates that radiation-induced reaction rates will be markedly different if the species are adsorbed on alumina, so one should regard the extrapolation of effects from a homogeneous to a heterogeneous system with some reservations about the predictions. Nevertheless, it is unlikely that oxidized species, such as MoO_4^{2-} and TcO_4^-, adsorbed on alumina would cease to be reduced by hydrated electrons and hydrogen atoms. The degree, and perhaps the rate, of reduction may be influenced by adsorption on alumina but not the likelihood of its occurrence.

Experience with the molybdate-pertechnetate generators using alumina adsorbent and sodium chloride eluant suggests that significant radiation-induced reduction occurs which gives rise to a nonelutable technetium species and consequent loss of elution efficiency. The hydrated electron, e_{aq}^-, has been suggested as the species responsible for the reduction. A logical treatment to prevent e_{aq}^- reduction is therefore the incorporation of e_{aq}^- scavengers in the eluant.

It is held by some[23] that the presence of liquid water within the system is detrimental to the efficient operation of the generator; this has led to the development of a 'dry' generator where residual eluant is removed from the alumina bed at the end of each elution cycle.

The results of Boyd and Matthews[24] cast doubt on the simplicity of this hypothesis. Although they confirmed that the removal of residual saline from the generator by displacement with air or nitrogen was highly effective, they also proved that liquid water was not detrimental since high efficiencies were maintained throughout the life of the generator when the residual saline was displaced with pure water.

The inference from these experiments was that the ^{99}Mo:^{99m}Tc generator system is sensitive to chloride ions. The hypothesis appeared to have been confirmed when experiments in which sodium sulfate or sodium perchlorate solutions were used in place of saline as the eluant, resulted again in the maintenance of high elution efficiencies. Subsequent experiments showed that the presence of dissolved organic carbon in the saline was also detrimental to elution efficiency.

Probably the least complicated remedy for low elution efficiency is modification of the alumina bed to include an insoluble scavenger of hydrated electrons. An early form of this method made use of the absorption of dichromate ions onto the alumina; more recently Panek-Finda[25] has described the use of alumina, coated with hydrated manganese dioxide (0.3% w/w Mn), which combines adequate capacity for molybdate absorption with sufficient oxidizing potential to overcome the reducing effects of radiolysis. Boyd and Matthews[26] report similar innovations with ceric oxide or silver nitrate coated alumina.

Technetium-99m generators rarely operate at 100% efficiency throughout their useful life. However, the manufacturer strives to achieve this goal via one (or a combination) of the techniques described here. In some cases, the efforts of the manufacturer to achieve maximum output have resulted in generators which reduce the quality of the separated ^{99m}Tc, e.g., low pH, elevated Al^{3+} contamination, the presence of oxidizing agents in the eluate. *Users should be prepared to monitor the performance of their generators; one objective method of doing this has been described in the literature by Boyd and Hetherington.*[19] The users should also be aware of the manufacturer's method of maximizing the ^{99m}Tc and what this means with respect to the eluate purity.

D. Radionuclidic Purity

Technetium-99m was selected for use in nuclear medicine because it entails minimum radiation exposure even when relatively high activities are administered. The presence of tracers of other radionuclides can increase radiation exposure to an unacceptable level and hence it is essential that the statutory limits on the nature and quantities of likely impurities are observed.

Unfortunately those impurity radionuclides which tend to be the most undesirable because of this ability to impart high radiation doses, i.e., the α- and β-emitters, are the most difficult to detect and quantify. Because they rarely have the resources to monitor radionuclidic contamination, users look to the manufacturers to maintain extremely high standards in their production processes. On the other hand, manufacturers are aware that the user can influence the radionuclidic content of the eluate by not adhering to the operating instructions.

The literature contains a number of reports of the radionuclidic impurities found in ^{99m}Tc eluted from chromatographic generators. In general, the nature of the impurities is dictated by the production route of the parent ^{99}Mo.

The radionuclidic quality of the eluates in generators produced from (n,γ) ^{99}Mo is governed by the purity of the original molybdenum target material. Elements present in trace quantities may undergo neutron activation along with the molybdenum and eventually appear as a radionuclidic contaminant in the separated ^{99m}Tc. The most likely impurity is, of course, ^{99}Mo which enters the eluate stream either by desorption or associated with fragments of the alumina bed.

However, after the ^{99}Mo has decayed to an insignificant level a range of long-lived radionuclidic impurities can also be identified. Meinhold et al.[27] have described two distinct types of (n,γ) ^{99}Mo generator; one type where the major long-lived impurity was ^{134}Cs, the other type where ^{110m}Ag predominated. They did not explain the apparent mutual exclusion of ^{134}Cs and ^{110m}Ag in generator eluates and their results indicate that both types of generators were made by a single manufacturer. The results of Finck and Mattson[28] point to a possible explanation. These workers showed that each of the radiocontaminants co-adsorbed with ^{99}Mo onto the alumina bed, shows a tendency to be eluted according to the volume of eluant passed through the column. Some radionuclides (e.g., ^{134}Cs) are rapidly eluted from the column and their concentration tends to decrease with increasing eluate volume — i.e., a wash-out process. Silver-110m, on the other hand, reaches a maximum concentration in the eluate only after a large volume has passed through the generator. It is possible, therefore, that Meinhold et al. obtained their results because of the different ways the generators were washed before elution; for those generators which had only limited washing, the ^{134}Cs content would have been relatively high; for those with extensive washing, the ^{134}Cs would have been removed and the ^{110m}Ag accentuated. Wood and Bowen[29] have reported finding ^{95}Zr and ^{124}Sb in the eluates from a (n,γ) ^{99}Mo generator; they also showed that these impurities could be removed by further washing of the column with saline.

Billinghurst and Hreczuch[30] emphasized the need for actual measurement rather than intuitive assumption when they detected fission products and neptunium-239 in the eluates of a generator loaded with (n,γ) ^{99}Mo. This observation suggests uranium contamination of the molybdenum target.

In generators produced with (n, fission) ^{99}Mo, other fission products and the transuranic alpha-emitting elements may be present and special attention must be given to both the processing and testing of such generators to ensure that these impurities are minimized. Besides ^{99}Mo contamination of the eluted ^{99m}Tc, the following gamma-emitting radionuclides may also be encountered; ^{131}I, ^{132}I, ^{111}Ag, ^{112}Ag, ^{112}Pd, ^{140}Ba, ^{140}La, ^{95}Zr, ^{95}Nb, ^{103}Ru and ^{137}Cs.

Of these, only ^{137}Cs shows an immediate decrease in concentration in a progressive series of elutions; this is presumably due to a rapid wash-out similar to that occurring with ^{134}Cs in the (n,γ) ^{99}Mo generator.

Silver-111 attains a maximum concentration (μCi per mCi ^{99m}Tc eluted) during the elution series, indicating the much slower elution of Ag species from alumina. (This also has been observed with the (n,γ) ^{99}Mo generator.) Ruthenium-103, ^{95}Zr/^{95}Nb, ^{140}Ba/^{140}La and ^{131}I all increase in concentration with increasing generator age; however, much of this change may be attributed to the decay of ^{99}Mo and the reduced production rate of ^{99m}Tc.

Silver-112 and ^{132}I are contaminants whose concentration can be influenced by the elution regime selected by the user, particularly with respect to the inter-elution period. Iodine-132 in the eluate is derived from ^{132}Te absorbed on the generator bed; similarly ^{112}Ag is derived from ^{112}Pd. The growth and quantity of elutable ^{132}I and ^{112}Ag is a function of time. Because both ^{112}Ag and ^{132}I have half-lives shorter than ^{99m}Tc, their rate of formation from the respective parent radionuclides is initially faster than that of ^{99m}Tc. As a result, an elution regime where the ^{99m}Tc is separated at frequent intervals will exhibit elevated values for the concentrations of ^{132}I and ^{112}Ag.

The resolution of a complex mixture of gamma-emitting radionuclides is achievable only with a high resolution Ge(Li) detector used in conjunction with a multichannel analyzer. Ideally, the spectral data are analyzed via a computer program[31] which identifies and quantifies the source of the peaks and calculates the time at which the level of each impurity exceeds a pre-set value (the expiry time).

Gamma-ray emitters can be easily detected by spectroscopy, but this is not so for the alpha and beta particle emitters and, as a result, generator users more commonly omit to measure this aspect of quality. Compliance with the limits defined in the pharmacopoeias is the goal sought by the manufacturers with their much greater resources of analysis.

Of the pure beta-emitting radionuclidic impurities likely to be encountered in the eluates from a fission-based ^{99}Mo:^{99m}Tc generator, only ^{89}Sr and ^{90}Sr are of prime concern; a method for determining ^{89}Sr and ^{90}Sr in generator eluates has been described by Sodd and Fortman[32] using the method of Velten.[33] This procedure is time consuming and subject to error because of incomplete gravimetric separation and contamination with other radionuclides. The sensitivity of the method is influenced by the background activity of the counting system but Sodd and Fortman claim a minimum detectable limit of 0.15 pCi. These workers showed that the eluates from the generators they examined contained ^{89}Sr and ^{90}Sr at levels which were at least three orders of magnitude less than the maximum levels considered acceptable.[34]

The most likely alpha-emitting impurities in eluates from a fission-based ^{99}Mo:^{99m}Tc generator are ^{235}U, ^{238}U, and ^{239}Pu; the presence of these radionuclides is attributable to both the uranium target and to the end product of neutron activation:

$$^{238}U \ (n,\gamma) \ ^{239}U \xrightarrow{\beta^-} \ ^{239}Np \xrightarrow{\beta^-} \ ^{239}Pu$$

Chemical processing to separate ^{99}Mo will substantially reduce the amounts of these impurities but the possibility remains that some will persist through to the eluted ^{99m}Tc.

Charlton[35] correlated the radiological risk with the maximum permissible concentration and suggested that the ^{239}Pu concentration should not exceed 0.4 pCi per mCi ^{99m}Tc. Sodd et al.[36] showed that by alpha spectrometric measurements on electrodeposits from generator eluates, ^{239}Pu could be determined with an extremely low sensitivity of 0.01 pCi per sample.

Fissile alpha-emitters can be detected with great sensitivity by the measurement of delayed neutrons emitted following neutron irradiation.[37] In this elegant technique, the sample is irradiated for approximately 1 min with thermal neutrons to induce ^{235}U and ^{239}Pu to undergo fission. After removal from the neutron beam, the sample is transferred immediately to a suitable counting array to measure the delayed neutrons arising from fission fragments. Even with counting times of less than 1 min, the minimum level of detection for ^{235}U is as low as 4×10^{-3} μg or 0.01 pCi per sample. Unfortunately, the minimum level of detection for ^{239}Pu is only 10^{-2} μg or 330 pCi per sample; hence use of this technique to monitor the ^{239}Pu content necessitates a sampling plan where the original ^{99m}Tc activity is high (> 330 mCi).

E. Radiochemical Purity

The elution of a generator which consists of ^{99}Mo as molybdate adsorbed on alumina results in the separation of ^{99m}Tc in predominantly the pertechnetate form. Although pertechnetate is the most stable valency state of technetium, lower valency species have, however, been detected in generator eluates. Vesely and Cifka[20] reported that a proportion of the generators they examined, presumably of an early design, yielded ^{99m}Tc with a nonpertechnetate component in excess of 10%. These workers also found that certain reduced species of technetium were strongly absorbed by alumina and that this effect was responsible for low elution efficiencies.

More recently designed ^{99}Mo:^{99m}Tc generators usually contain devices which protect the pertechnetate species from radiolytically induced reduction reactions and it is now most unusual for the nonpertechnetate species to exceed 5% of the total eluted ^{99m}Tc.

Analytical techniques to separate Tc(IV), Tc(V), and Tc(VII) have been described by Shukla[38] and, more recently, by Shen et al.[39] Kits using miniaturized thin layer sheets of silica gel with various solvent systems, provide a rapid and accurate assessment of the radiochemical species of ^{99m}Tc contained in generator eluates.

F. Chemical Purity

Eluted ^{99m}Tc may contain certain chemical impurities, originating from either the generator bed or the eluant, which could detrimentally affect the clinical application of the radionuclide.

Probably the chemical impurity most commonly (albeit ambivalently) reported is aluminum. Aluminum cations are formed during absorption of the ^{99}Mo when the alumina bed is subjected to a strongly acid environment. Although subsequent washing of the generator removes almost all the Al^{3+} ions, a residue, whose quantity depends on the extent of the washing process, can still be detected.

Weinstein and Smoak[40] reported that the in vitro labeling of erythrocytes with ^{99m}Tc was affected by Al^{3+} present in generator eluates; they observed that agglutination of the red cells also occurred with Al^{3+} concentrations as low as 5 µg per mℓ. This was confirmed by Lin et al.[41] who also suggested a plausible mechanism for the effect. They postulated that, since the effect was observable only at low pH values, clumping was caused by the formation of cationic Al^{3+} bridges between the negatively charged red blood cells. They concluded that because the necessary conditions did not exist in vivo, intravascular agglutination following administration of an Al-containing generator eluate was highly improbable.

Other workers[42,43] have suggested that Al-contamination affects the quality of certain radiopharmaceuticals prepared from generator eluates and have recommended the use of a sequestering agent, such as EDTA, to overcome the problem of inconsistent quality. Craigin et al.[44,45] found Al-contamination levels in generator eluates ranging from nil to 40 µg/mℓ but could prove no correlation between aluminum content and the clinical performance of batches of ^{99m}Tc-sulfur colloid made from the eluates. Chaudhuri[46,47] reported an altered biological distribution of ^{99m}Tc-disphosphonate prepared in the presence of 50 µg aluminum ions. More recently Shukla et al.[48] have suggested that the presence of as little as 4 µg/mℓ aluminum ions changes the pertechnetate species to a 'pertechnetatoaluminum' complex. These investigators provided evidence of altered chromatography and biodistribution of the pertechnetate species in the presence of aluminum ions.

An explanation of the conflicting reports in the literature on the detrimental effects of contaminating Al^{3+} ions may be found in the final observation of Shukla et al. They reported that the 'pertechnetatoaluminum' complex underwent hydrolysis on standing; their evidence suggested that the pertechnetate ion was reliberated within 4 hr, after which normal biodistribution returned.

Undesirable chemical impurities may also be imparted to the ^{99m}Tc solution from the eluant. Oxidizing agents (e.g., hypochlorite, hydrogen peroxide, dichromate) added to maximize ^{99m}Tc yields will interfere with the chemical reactions which involve Sn(II) reduction of the pertechnetate ion and which are used subsequently in the preparation of radiopharmaceuticals.

The use of scavenging agents (e.g., nitrates) for radiolytically induced hydrated electrons must also be viewed as potentially likely to complicate subsequent chemical reactions.

An often overlooked impurity which can seriously affect the chemical and biological properties of eluted ^{99m}Tc and radiopharmaceuticals prepared from it, is dissolved organic carbon originating from the plastic components of the generator. Saline enclosed within a vinyl sachet is contaminated with plasticizers (e.g., di-iso octyl phthalate)[49] which may be reactive towards ^{99m}Tc and compete with other ligands in the presence of a reducing agent to give lowered labeling yields.

In the modern designs, the combined goals of enhanced yield from the generator and suitability of the eluted ^{99m}Tc for subsequent incorporation into a range of radiopharmaceutical preparations are achieved without recourse to chemical additives in the eluant. Hence, the problems of maintenance of chemical purity and avoidance of potential toxicity reduces the selection of appropriate construction materials for the total generator system and adherence to good manufacturing practice.

G. Biological Purity

Since the ^{99m}Tc eluted from a generator will normally be used directly or, following chemical manipulation as an intravenous injection, attention must be paid to sterility and freedom from pyrogens.

Charlton[50] drew attention to the fact that the generator system presented a difficulty additional to those normally prevailing with pharmaceutical products, in that the final control of quality is not in the hands of the manufacturer but of the user. This has led the manufacturers to design generators which, after being autoclaved at the factory, can be operated according to techniques which avoid further bacterial contamination. Sorensen et al.[51] have shown that the chromatographic generator effectively filters microorganisms and that eluates from manufactured sterile generators do not require terminal autoclaving as a further safeguard before injection. However, other work[52] has clearly indicated that the alumina bed of the generator cannot be relied upon as a fully effective bacterial filter and so eliminate even the low levels of contamination encountered during manufacture.

The radiation field, inherent to the ^{99}Mo:^{99m}Tc generator system and surrounding the alumina column, is also no guarantee of eluate sterility since the source of the contamination may be the generator pipework which is relatively remote from the effects of the contained radioactivity.

In summary, therefore, the manufacturer will maintain high standards of biological purity by

1. *Effectively housing the manufacturing facility to minimize bacterial access*
2. *Thoroughly cleansing and sterilizing the generator components*
3. *Terminally sterilizing the loaded generator (e.g., by heat or external radiation)*
4. *Providing a means by which the generator may be aseptically eluted*

Adoption of these precautions eliminates the need for the user either to sterilize or to test the generator eluates for sterility.

IV. THE SUBLIMATION GENERATOR

In 1937 Perrier and Segre,[1] the discoverers of technetium, reported the volatility of the oxide of technetium (Tc_2O_7) and demonstrated that the technique of sublimation could be used to separate technetium (and rhenium) from molybdenum. Morgan and Sizeland[53] used this characteristic to produce trace amounts of ^{99m}Tc from irradiated molybdenum trioxide.

Robson and Boyd[54] predicted that the different volatilities of molybdenum trioxide and technetium heptoxide could be exploited to provide a source of medically acceptable ^{99m}Tc. Robson[55] and Lee[56] subsequently developed a practical sublimation generator able to handle 200 g of low specific activity (1 Ci/g Mo) neutron irradiated molybdenum trioxide. This is shown diagrammatically in Figure 5.

In this apparatus the irradiated molybdenum trioxide is loaded into a horizontal tube furnace through which is flowing a stream of oxygen; the temperature is raised to 850°C at which the ^{99m}Tc sublimes and is carried from the furnace to a condenser by the oxygen stream. Some of the molybdenum trioxide is volatilized at this temperature and is removed from the gas stream by a porous plug which is maintained at a temperature between that of the furnace and the boiling point of Tc_2O_7 (310°C). The vapor of the ^{99m}Tc compound enters the condenser after passing through the porous plug and condenses on a cooled surface.

Robson speculated that, because the sublimed deposit of ^{99m}Tc dissolved readily in pure water to yield pertechnetate, the principal technetium compound involved in the sublimation process was probably the oxide Tc_2O_7.

Tachimori et al.[57] studied the separation of ^{99m}Tc from neutron-irradiated molybdenum trioxide at elevated temperatures and found evidence to suggest that the ^{99m}Tc was released in the form of the dioxide, TcO_2, which then oxidized to Tc_2O_7 only when the procedure

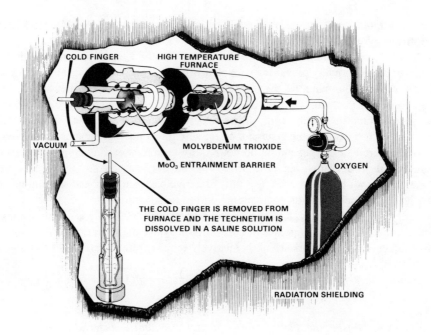

FIGURE 5. Diagrammatic view of the sublimation generator.

was carried out under an oxidizing atmosphere. They found that working at temperatures too low to volatilize molybdenum trioxide, the separation of ^{99m}Tc was governed by a bulk diffusion mechanism which gave yields too poor to be of practical importance. Their more fundamental studies of the process confirmed the pragmatic approach of both Robson and Lee by concluding that a practical sublimation generator must involve some co-volatilization (and subsequent condensation) of the molybdenum trioxide.

In an extensive series of articles on the thermal separation of ^{99m}Tc from molybdenum trioxide, Vlcek et al.[58-62] confirmed many of the earlier findings on the performance characteristics of the sublimation generator. These characteristics are summarized below:

1. The yield of ^{99m}Tc diminishes markedly on repeated sublimation
2. The yield can be enhanced by raising the furnace temperature
3. More constant ^{99m}Tc yields can be obtained by the stepwise increase of temperature with each successive sublimation
4. Optimal operation occurs at temperatures where the molybdenum trioxide also sublimes
5. Because the mechanism of the release of ^{99m}Tc from MoO_3 is bulk diffusion, the yield of ^{99m}Tc can be maximized by increasing the dwell-time at sublimation temperature
6. The yield of ^{99m}Tc is also influenced by the mass of the MoO_3 loaded into the sublimation furnace

Mass MoO_3 (g)	Temperature (°C)	Dwell-time (min)	Separation efficiency (%)	Source
<1	800	30	>80	Tachimori et al.
10	830-900	60	80	Vlcek et al.
200	850	100	40	Lee

A. Radionuclidic Purity of Sublimed ^{99m}Tc

In the case of the sublimation generator, the radionuclides, besides ^{99m}Tc, which appear in the product are those which have a significant vapor pressure at the temperature of the furnace. Under normal operating conditions the sublimation generator provides ^{99m}Tc in which the ^{99}Mo levels are consistently held in the range 10^{-3} to 10^{-4} per cent of the total radioactivity. If the maximum permissible concentration of ^{99}Mo in ^{99m}Tc is taken as 0.1% and the expression for expiry time as

$$(T_{exp})_{Mo} = \frac{1}{\lambda_{Tc} - \lambda_{Mo}} \cdot \ln\left(\frac{0.1}{x}\right) \tag{8}$$

where x = measured per cent ^{90}Mo in ^{99m}Tc, then ^{99m}Tc produced by the sublimation technique is shown to be suitable for use, in terms of ^{99}Mo contamination, for 44 to 66 hr postseparation.

Other radiocontaminants found in sublimed ^{99m}Tc originate from trace chemical impurities in the molybdenum trioxide target material, principally rhenium, tungsten, and uranium.

Rhenium impurities undergo neutron activation to form ^{186}Re and ^{188}Re via the nuclear reactions

$$^{185}Re\ (n,\gamma)\ ^{186}Re;\ T_{0.5} = 90.6\ \text{hours}$$

$$^{187}Re\ (n,\gamma)\ ^{188}Re;\ T_{0.5} = 17.0\ \text{hours}$$

The chemical similarity of rhenium to technetium ensures that the rhenium radionuclides formed during neutron irradiation eventually collect in the separated ^{99m}Tc. It is difficult, however, to determine accurately low levels of the radiorheniums in ^{99m}Tc because of either a similarity in energy of the emitted photons (^{186}Re; 0.137 MeV — 9.2% disintegrations: ^{188}Re; 0.155 MeV — 15% disintegrations) or the actual paucity of higher energy photons (^{188}Re; most abundant photopeak after 0.155 MeV is 0.633 MeV — 1.4% disintegrations). Rhenium-188 alone, may be determined by gamma spectrometry if the contribution to the spectrum from ^{99m}Tc is reduced by screening the sample with a few millimeters of lead; in these circumstances the photopeaks at 0.478 and 0.633 MeV are sufficiently unique to ^{188}Re to permit quantification. However, for the ^{188}Re determination to be used also as a measure of the ^{186}Re contamination a correction factor must be applied to account for the different rates of decay of ^{186}Re and ^{188}Re. The situation is further complicated by the presence of tungsten impurities in the molybdenum trioxide; when neutron activated, tungsten produces ^{188}Re according to the reaction:

$$^{186}W\ (n,\gamma)\ ^{187}W(n,\gamma)\ ^{188}W \xrightarrow{\beta^-} \ ^{188}Re$$

Although the yield from the double-capture reaction to form ^{188}W is small, it does represent a regenerative and long-lived source of ^{188}Re. As the levels of ^{188}Re and ^{186}Re, formed by the activation of rhenium impurities in the molybdenum trioxide, are depleted by the combined effects of radioactive decay and sublimation, so the small contribution of ^{188}Re from the decay of ^{188}W assumes greater significance. The actual value of the ^{186}Re:^{188}Re ratio depends upon such conditions as:

- The concentration of rhenium in the MoO_3
- The concentration of tungsten in the MoO_3
- The postirradiation age of the MoO_3 at the time of separating the rhenium radionuclides
- The number of times the rhenium radionuclides are separated
- The frequency at which the separation process is performed
- The efficiency of the separation process
- The flux of the irradiating neutrons

Radiorhenium levels generally fall away with the increasing number of sublimations of ^{99m}Tc out of the irradiated molybdenum trioxide. Typical results are

First sublimate	$4.3 \times 10^{-3}\%$ ^{188}Re	(measured)
	$2.6 \times 10^{-3}\%$ ^{186}Re	(calculated)
Seventh sublimate	$6.6 \times 10^{-5}\%$ ^{188}Re	(measured)
	$8.5 \times 10^{-5}\%$ ^{186}Re	(calculated)

As before, by adopting the maximum permissible concentration of the radiorheniums in ^{99m}Tc as 0.01%, and using the expression for expiry time:

$$(T_{exp})_{Re} = \frac{1}{2\lambda_{Tc} - \lambda_{188} - \lambda_{186}} \cdot \ln\left(\frac{0.01}{Y \times Z}\right) \qquad (9)$$

where Y = % ^{188}Re in ^{99m}Tc, Z = % ^{186}Re in ^{99m}Tc, the radiorhenium content of these sublimates can be shown to restrict the useful life of the ^{99m}Tc to 37 hr (first separation) and 79 hr (seventh separation).

The presence of uranium impurities in the molybdenum trioxide is indicated by detectable levels of volatile fission products in the sublimed ^{99m}Tc. Vlcek et al. reported the presence of ^{103}Ru which rapidly decreased to $10^{-4}\%$ of the ^{99m}Tc activity. Another fission product very occasionally detected is ^{132}I; because ^{132}I is produced by the decay of ^{132}Te ($T_{0.5}$ = 77.9 hr) its occurrence is likely to be observed in somewhat constant concentrations throughout the total life cycle of the irradiated molybdenum trioxide.

In general, the radionuclidic purity of sublimed ^{99m}Tc is extremely high and gives rise to expiry times in excess of 30 hr. This is indeed fortunate because the sublimation generator has been successfully exploited in a manufacturing environment only where the lag time between preparation and use often approaches 24 hr.

Colombetti et al.[63] developed a small sublimation generator which was designed for exploitation in a clinical laboratory using carrier-free (n,fission) ^{99}Mo. This device is reported to operate at high efficiencies (70 to 80%) and to produce ^{99m}Tc with levels of radionuclidic impurities lower than those found in the eluates of chromatographic generators, and undetectable levels of chemical impurities.

B. Other Purity Aspects Associated with Sublimed ^{99m}Tc

Since ^{99m}Tc is separated from MoO_3 at high temperatures and in the presence of oxygen, the radiochemical form of sublimed ^{99m}Tc is likely to be pertechnetate (Tc VII). The simplicity of the sublimation system and the absence of other reactive species, particularly of an organic nature, lessens the likelihood of reduced or complexed ^{99m}Tc species being formed.

Chemical impurities are low, particularly if the molybdenum trioxide is calcined to remove volatiles before irradiation. At the temperature of sublimation (850°C) molybdenum trioxide is extremely corrosive to containment materials. Some care in equipment design and maintenance is necessary to prevent corrosion products being entrained which will contaminate the ^{99m}Tc sublimate. The small amount of molybdenum trioxide vapor which passes through the entrainment barrier and condenses with the ^{99m}Tc, does not readily dissolve in a cold saline solution.

Freedom from microbiological impurities is assured with little special precaution beyond maintaining the appropriate standards of cleanliness for the condenser.

V. THE SOLVENT EXTRACTION GENERATOR

The solvent extraction generator has been widely exploited and considerable use made of its inherent advantage of economics — through the application of relatively cheap (n,γ)

^{99}Mo — and the technical advantage it has of providing a high concentration of ^{99m}Tc with low levels of radionuclidic impurities.

Gerlit[64] first reported the potential of organic solvents for the separation of technetium and molybdenum. In each case where technetium was extracted by the solvent, Gerlit found that rhenium was also extracted; this is another example of the very close similarity between the two elements. Allen[65] showed that when a column of alumina containing absorbed ^{99}Mo was eluted with methyl ethyl ketone, ^{99m}Tc was stripped at high efficiency; this method was recommended as a means of avoiding the radionuclidic and chemical impurities present in the eluates of the early chromatographic generators. A modification of this technique was also applied by Harper et al.[66] to purify generator eluates.

Various workers[67,72] have described laboratory techniques for the extraction of ^{99m}Tc from alkaline solutions of sodium molybdate (^{99}Mo) using methyl ethyl ketone (MEK). All used the same basic technique which involved

1. Bringing the aqueous ^{99}Mo solution into vigorous contact with a volume of MEK
2. Allowing the phases to separate when the less dense MEK forms the upper layer
3. Retaining the ^{99}Mo solution for subsequent re-extraction
4. Passing the MEK solution of ^{99m}Tc through a small alumina column to remove any traces of ^{99}Mo
5. Evaporating the MEK solution to dryness in a stream of warm air
6. Redissolving the ^{99m}Tc residue in a small volume of physiological saline

The practical problems associated with these apparently simple chemical operations were significant because radiation safety considerations dictated that the procedure be performed under remote handling conditions.

The critical steps in the separation process were the mixing of the aqueous and organic phases to ensure adequate contact and then, after the phases had settled out, the locating of the liquid:liquid interface so that the MEK solution of ^{99m}Tc could be removed. As a result of these difficulties, the solvent extraction generator developed into a complex device whose operation required highly skilled personnel.

Tachimori et al.[72] have reported an interesting variation on the basic process where they avoided the somewhat hazardous operation of MEK evaporation. These workers showed that ^{99m}Tc can be directly recovered from the organic phase by adsorption onto an alumina column followed by elution with saline.

Other methods of separating ^{99m}Tc from ^{99}Mo by solvent extraction techniques have been reported,[73-75] however, none appears to be simpler than the basic MEK process. Toren and Powell[76] developed a solvent extraction generator for which the loading with ^{99}Mo solution, the extraction cycles and finally the unloading of the extractor were programmed electronically. Charlier et al.[77] have also described an automatic solvent extraction generator which gave encouraging results.

The use of the solvent extraction technique for the large-scale manufacture of ^{99m}Tc has been reported by Sorby and Boyd.[78] The Australian radiopharmacy service at the AAEC Research Establishment, Lucas Heights, has largely relied upon this method of production to satisfy a national demand. The AAEC extraction plant can handle 250 g molybdenum trioxide at an initial specific activity in excess of 1 Ci ^{99}Mo per g MoO_3.

In the design of the apparatus the following facts, pertaining to the conditions of extraction, were recognized:

- Potassium molybdate is more soluble in potassium hydroxide than are the sodium equivalents
- The value of the distribution coefficient for ^{99m}Tc in MEK is influenced by the molybdenum concentration

- The separation of the phases occurs more rapidly at lower alkali concentrations

To avoid the problems of radiolytic reduction, common in concentrated solutions of (n,γ) ^{99}Mo and giving rise to a black precipitate, the Australian apparatus holds the molybdenum solution between extractions at 90 to 95°C with air bubbling; no other oxidants are added.

The apparatus, constructed primarily of stainless steel components, is undoubtedly complex and requires a high degree of instrumental control. Nevertheless, it has proved to be operationally reliable for several years under a regime of continual use at a rate of three extractions per day.

A. The Efficiency of the Solvent Extraction Generator

The solvent extraction generator's overall efficiency to produce ^{99m}Tc reflects the multiple stages of the total process and how effectively each stage can be accomplished.

Efficient solvent extraction will depend upon such design parameters as the ability to provide intimate and extended contact between the immiscible phases, the concentration of dissolved molybdate and the alkalinity of the aqueous phase. However, the extraction efficiency is also influenced by the radiochemical form of the ^{99m}Tc; Gerlit[64] discovered that technetium in a reduced form is not extracted by ketones; hence, steps must be taken to maintain the ^{99m}Tc in the heptavalent form. The tendency for ^{99m}Tc to undergo radiolytic reduction is commonly overcome by adding oxidizing agents to the aqueous phase, e.g., H_2O_2, $NaOCl, O_2$. Baker[71] has reported that the separation efficiency is influenced by the specific activity of the ^{99}Mo and that lowering the specific activity causes a significant drop in the average separation efficiency. Evidence which conflicts with this has been derived from almost 500 loadings of the AAEC's high activity apparatus; here the records show a weak but nevertheless significant correlation between average separation efficiency and initial activity of the ^{99}Mo charge, in that there was a 10% drop in efficiency for every 100 curies ^{99}Mo loaded into the apparatus.

The next stage of the process, the separation of the phases after contact, can also be a source of ^{99m}Tc loss which must be minimized by good design and careful operation. Although it is almost impossible to separate the phases exactly on the interface, the apparatus should allow the operator to approach this ideal. The use of a large volume of MEK minimizes phase-separation loss.

In the final stage, it is important that the ^{99m}Tc is recovered efficiently from the evaporator after the MEK has been removed. The presence of resinous material is definitely detrimental to ^{99m}Tc recovery and, in severe cases, can prevent a significant proportion of the radionuclide dissolving in the saline rinse.

B. Radionuclidic Purity of Solvent-Extracted ^{99m}Tc

The most common radionuclidic impurities in solvent-extracted ^{99m}Tc are ^{99}Mo, ^{186}Re, and ^{188}Re. Depending on the quality of the molybdenum target used and the adequacy of the extraction-separation process, the levels of these impurities are often vanishingly small ($10^{-5} - 10^{-4}$%).

The radiorheniums originate from Re and W impurities in the molybdenum target and their relative proportions are governed by exactly the same complex set of conditions of production and separation as were described for the sublimation generator. In general, the radiorheniums are present in higher concentrations at the beginning of a series of extractions and then rapidly disappear until a roughly equilibrium situation is achieved, corresponding to the regenerating growth of ^{188}Re from ^{188}W.

On the other hand, the risk that ^{99}Mo will contaminate the ^{99m}Tc is ever present and represents a monitor on the functioning of the separation process. Incomplete phase-separation will cause ^{99}Mo-contamination of the MEK evaporator and subsequent ^{99m}Tc product solutions. Fortunately, this can be prevented by passing the MEK stream and/or the ^{99m}Tc saline solution through a small alumina column which selectively removes ^{99}Mo. The low specific activity of the ^{99}Mo combined with the limited absorptive capacity of alumina for molybdenum demand that the alumina filter be changed regularly; an unusually high level of ^{99}Mo in ^{99m}Tc indicates the need to replace the alumina.

Several long-lived radionuclidic contaminants have also been detected in solvent-extracted ^{99m}Tc. These are caused by impurities in the molybdenum target.

C. Chemical Purity of Solvent-Extracted ^{99m}Tc

The use of methyl ethyl ketone of inadequate quality will have a detrimental effect on the purity of extracted ^{99m}Tc because impurities are concentrated during the evaporation process. It is necessary, therefore, to redistill the MEK before use to avoid the gummy residue that may be left after evaporation. However, Hunter[79] has commented that even when reagents of the highest available quality have been employed, aldol condensation products were formed immediately and subsequently interfered in labeling reactions involving a stannous reduction. Boyd found that the addition of stannous chloride to an aqueous solution of MEK-derived ^{99m}Tc gave rise to a bright yellow color which quickly faded on standing. Baker[71] reported that if air instead of nitrogen was used in the evaporation of the MEK, the result was a yellow solution of pertechnetate, presumably because the ketone is polymerized.

Narasimhan and Mani[80] found direct evidence to suggest that, in alkaline solutions of MEK, polymeric condensation products are formed which are extracted (along with the ^{99m}Tc) by the MEK solvent. After the MEK has evaporated, the condensation products form a nonvolatile residue which partially dissolves in the saline to contaminate the ^{99m}Tc. These workers found that acidification of the ^{99m}Tc in saline solution produced a brown precipitate. Their experiments also indicated that if the MEK-^{99m}Tc stream were passed through an alumina column before evaporation, the condensation products were removed and the ^{99m}Tc subsequently produced was of a high quality and suitable for use in nuclear medicine in the form of various labeled compounds.

Despite the assurance that high quality ^{99m}Tc may be achieved so easily, comments in the literature[81,82] to the effect that the use of MEK-derived ^{99m}Tc can lead to poor labeling efficiencies, stand largely unrefuted. More recent experience has indicated that the in vitro labeling of red blood cells is also sensitive to ^{99m}Tc from different sources; poor labeling efficiency is attributed to, inter alia, the use of MEK-derived ^{99m}Tc. The biological property of the phosphorus-containing bone-seeking radiopharmaceuticals may be altered when MEK-derived ^{99m}Tc is incorporated into the formulation; an increase in the proportion of the ^{99m}Tc retained in the vascular compartment being the most common observation.

The chemical purity of MEK-derived ^{99m}Tc may be improved by electrolysis. In this technique, the ^{99m}Tc is removed from the saline solution containing the organic residues by electrodeposition onto a platinum cathode and subsequently redissolved in fresh saline.

VI. COMPARISON OF ^{99m}Tc-GENERATOR FORMS

Chromatographic Generator Containing (n,γ) ^{99}Mo Produced ex Natural Molybdenum

Advantages	Simple processing.
	Only inexpensive low-specific activity ^{99}Mo required.
	Yield not limited by extensive waste disposal problems.
	Simple to operate and portable.
	^{99m}Tc separated with high efficiency.
Disadvantages	Large size alumina column required for low specific activity ^{99}Mo; low elution profile — large eluate volume.
Future prospects	Development of substrates other than alumina required to reduce the physical dimensions of the generator.[83]

Chromatographic Generator Containing (n,γ) ^{99}Mo Produced ex Enriched ^{98}Mo

Advantages	Simple processing of ^{99}Mo.
	Easy to operate and portable.
	^{99m}Tc separated with high efficiency.
	Elution profile improved.
Disadvantages	Expensive target material and a very high flux reactor required for maximum effect; probably the most expensive method of producing ^{99m}Tc unless the ^{98}Mo is recycled.
Future Prospects	Obsolescence.

Chromatographic Generator Containing Fission Produced ^{99}Mo

Advantages	Use of physically small generators permitted because of carrier-free characteristics of ^{99}Mo; giving not only maximum radioactive concentration of ^{99m}Tc but also reduced shielding mass.
	Simple to operate and portable.
	^{99m}Tc separated with high efficiency.
	Excellent elution profile.
Disadvantages	High capital cost of processing plant; special problems arising from gaseous fission products and subsequent environmental hazards.
	Elaborate processing precautions required to avoid fission product or transuranic contamination of ^{99}Mo.
	Disposal problem of other fission products.
	High cost per mCi ^{99m}Tc.
Future prospects	Although this type of generator is currently enjoying widespread usage, the inherent disadvantages are encouraging the search for alternatives. It is predicted therefore, that the fission produced ^{99}Mo generator will eventually be superseded by advanced designs incorporating (n,γ) ^{99}Mo.

The Sublimation Generator Containing (n,γ) ^{99}Mo Produced ex Natural Molybdenum

Advantages	Only inexpensive low-specific activity ^{99}Mo required. No chemical processing required.
	Capable of being scaled up to kilocurie quantities.
	Product free from chemical impurities.
	Very high radioactive concentrations easily attained.
	High radionuclidic purity.
	Low cost per mCi ^{99m}Tc.
Disadvantages	Version suitable for use in small nuclear medicine laboratories yet to be developed.
	Separation efficiency reduced (25 to 50%) and specially programmed process cycles required to prevent even further efficiency deterioration.
Future prospects	Despite its ability to produce ^{99m}Tc of exceptionally high quality, it is unable to be fully exploited until quantitative performance characteristics have been improved. Further development required.

The Solvent Extraction Generator Containing (n,γ) ^{99}Mo Produced ex Natural Molybdenum

Advantages	Only inexpensive low specific activity ^{99}Mo required.
	Capable of being scaled up or down to individual requirements.
	High radioactive concentration ^{99m}Tc attainable.
	High radionuclidic purity.
	High separation efficiency.
	Low cost per mCi ^{99m}Tc.
Disadvantages	Apparatus complicated; highly trained personnel required.
	Possible fire hazard with MEK vapor.
	Possible interference by polymeric organic residues in the ^{99m}Tc product solution with subsequent tagging reactions to produce undesirable changes in biological properties.
Future prospects	Only limited if the problem of organic residues, which has brought this method into disrepute, is not resolved.

VII. RADIOPHARMACEUTICALS OF ^{99m}Tc

Considerable research has been invested in the development of compounds which combine organ specificity with the recognized physical advantages of ^{99m}Tc; the literature of the past two decades has been deluged with descriptions of the biological activity of numerous ^{99m}Tc formulations and the search for new agents continues. No attempt is made here to discuss the advantages or limitations of each ^{99m}Tc-compound that has been proposed, even though some of them represented a significant step advance in the development of what is essentially a new technology.

The early development of ^{99m}Tc-radiopharmaceuticals took place within the university-linked nuclear medicine clinics and national laboratories which seek to extend nuclear science beyond energy production. As a result, the first generation of ^{99m}Tc-agents frequently required synthesis *in situ;* these agents often had a span of usefulness which was limited by considerations of physical half-life and radiation-induced decomposition. However, as the techniques of organ imaging took hold and became an integral part of the process of diagnosis, the interest of the commercial drug houses was aroused, these limitations were overcome and the development of more durable kits ensued. These kits contained all the essential nonradioactive reagents in a measured and convenient form (e.g., solution, freeze-dried pellet, etc.); there remained the need simply to add ^{99m}Tc to the mixed reagents to produce a highly specific organ-seeking radiopharmaceutical. This is now the usual manner in which ^{99m}Tc is used within a routine nuclear medicine clinic.

A. Technetium Chemistry

Technetium occupies a position between manganese and rhenium in Group VII of the periodic table and is characterized by having multiple valence states. The most stable state is (+7) as for example in the pertechnetate ion in which form it separates from ^{99}Mo in the various generator options. Pertechnetate will act as an oxidizing agent and hence the most probable reactions for this anion involve reduction to a lower oxidation state. Reduction normally results in a cationic technetium species, such as $TcO(OH)^+$ which may hydrolyse and be precipitated as TcO_2 or, alternatively be stabilized in solution by chelation. Most technetium-based radiopharmaceuticals contain the radionuclide in a reduced form[84] using a complexing ligand to direct the biological behavior.

In their excellent review, Eckelman and Levenson[85] list the wide variety of reducing agents that have been used in the preparation of ^{99m}Tc radiopharmaceuticals; the most common has been tin (II), but others include iron (II), iron (III) with ascorbic acid, sodium borohydride, concentrated hydrochloric acid, and thiosulfate in acid medium. More recently, titanium (III), copper with ascorbic acid, and formamidine sulfinic acid have been added to the list of potentially useful reducing agents.[86-88]

The oxidation state and ionic charge of reduced ^{99m}Tc are difficult to obtain by direct measurement because of the very low concentration of ^{99m}Tc solutions. A stratagem that has been adopted is the measurement of these parameters for solutions of the longer-lived ^{99}Tc where the concentration of technetium is several orders of magnitude higher. Despite the uncertainties[89] involved in extrapolating these findings to the analogous situation with ^{99m}Tc, an understanding is evolving of the effect of reducing agent, ligand and pH on the ^{99m}Tc species.[90-95] Generalizations put forward by Russell and Cash[94] assert that the stable oxidation states of technetium complexes in aqueous media are III, IV, and V. Stability depends upon pH; Tc(IV) is found at all pH values, whereas Tc (III) is more characteristic of acidic media and Tc(V) of alkaline. More than one oxidation state may co-exist at any given pH.

Significant use has also been made of electrochemical reactions of ^{99m}Tc. Benjamin[96] pioneered the technique of labeling human serum albumin with technetium by electrolysis in a cell containing a zirconium anode. Conflicting theories[97,98] have been put forward to

Table 1
COMMON ^{99m}Tc RADIOPHARMACEUTICALS

Compound, chelate	Organ(s) visualized
Pertechnetate solution	Brain; thyroid
Colloidal dispersions based upon: Tc/S, Tc/Phytate (Ca), Tc/SnO$_2$, Tc/Sb$_2$S$_3$	Liver, spleen and bone marrow; lung (inhalation); lymph nodes
Large particle (15-50 μm) suspensions based upon: Tc/Fe(OH)$_2$, Tc/Albumin	Lung (perfusion)
Solutions of Tc/DTPA; Tc/Gluconate; Tc/Heptagluconate; Tc/Citrate; Tc/Dimercaptosuccinic acid	Kidneys; brain; tumors (some)
Tc-Protein complexes: albumin, fibrinogen, urokinase, streptokinase	Blood pool; deep vein thrombosis
Tc-P complexes: Poly- and pyrophosphate; diphosphonate	Skeleton; myocardial infarction
Tc-Iminodiacetic acid derivatives	Liver, bile ducts, gall bladder
Tc-Blood Cells; erythrocytes	Heart, spleen

explain the mechanism of the reaction; the weight of the evidence appears to support the more familiar concept of chelation by the ligand after the pertchnetate has been reduced to a lower oxidation state. The chelation of ^{99m}Tc via electrolysis using tin electrodes[99-101] has been attributed to the same mechanism. More recent work by Russell and Majerik[102] has shown that provided the electrode potential is controlled within certain limits and in the presence of a chelating agent, ^{99m}Tc can also be reduced at inert electrodes.

The clinical application of many of the ^{99m}Tc compounds has often preceded the elucidation of the underlying chemistry. The short half-life of ^{99m}Tc, its radioactivity and extremely low concentration (nanomolar) are complications which have encouraged a rather empirical approach; nevertheless, remarkable success has already been achieved in the absence of an understanding of the precise nature of technetium compounds and the relationship between structure and biologic activity. Despite the past successes, however, Eckelman and Levenson have predicted that, in the further development of ^{99m}Tc radiopharmaceuticals, there is a prerequisite to resolve the remaining uncertainties in its chemistry.

B. Technetium Radiopharmaceuticals

Many substances, containing ^{99m}Tc, have been designed with the appropriate biological properties to serve as diagnostic radiopharmaceuticals. A selection of these, including their region of use is given in Table 1.

The 6 hr half-life of ^{99m}Tc, while contributing significantly to reducing the patient's exposure to radiation, is disadvantageous in terms of supply logistics, and initially required the user to be involved in the synthesis of ^{99m}Tc-radiopharmaceuticals. This activity was not without its problems since variation in techniques and unsuspected radiation-induced decomposition often imparted variable qualities to the hospital-produced radiopharmaceutical. Although a more rigid control of procedures minimized quality deficiencies, the situation was not ideal because the production of high quality radiopharmaceuticals demanded facilities and expertise not commonly found in any but the largest hospitals. Two developments in the supply of radiopharmaceuticals have diminished this problem.

The first involved removing the need for the clinical department to perform the sometimes complex chemical manipulations on the rapidly decaying ^{99m}Tc. This was achieved by separating the radiopharmaceutical into its two component parts, i.e., the radionuclide and the substances which act upon it to bestow biological activity. It was possible to preconstitute the nonradioactive components (e.g., complexing agent, pH adjustment additives, reducing agent, stabilizers, etc.) in metered amounts suitable for unit or multiple doses in a sealed container. From this device a sophisticated radiopharmaceutical is available at any time simply by the addition of pertechnetate, the form that ^{99m}Tc assumes on separation from a generator.

The second development which furthered the practical exploitation of ^{99m}Tc was the creation of the centralized radiopharmacy. Whether on a metropolitan, regional, or national basis, the centralized radiopharmacy can make significant contributions to patient-care by making high quality, short-lived radiopharmaceuticals readily available. The centralized radiopharmacy creates a pool of expertise which can manage the resources of ^{99m}Tc with maximum effect, prepare the various reagent kits, perform comprehensive testing schedules, and, finally, arrange delivery to the user hospitals.

The modern history of ^{99m}Tc has been compressed into the remarkably short time of 20 years. It has become the radionuclide of choice because of its physical properties and ^{99m}Tc-radiopharmaceuticals have proved to be very versatile and have almost excluded all its competitors. The development of advanced generators of ^{99m}Tc together with the complementary reagent kits have ensured that these benefits can be universally enjoyed.

A problem which is now assuming more significance is the lack of an adequate knowledge of the chemistry of ^{99m}Tc. Several investigators see this deficiency as a limiting factor in the further development of ^{99m}Tc but regard it as a fertile area for future research.

REFERENCES

1. **Perier, C. and Segre, E.**, Radioactive isotopes of element 43, *Nature (London)*, 140, 193, 1937.
2. **Perrier, C. and Segre, E.**, Some chemical properties of element 43, *J. Chem. Phys.*, 5, 712, 1937.
3. **Erdtman, G. and Soyka, W.**, The gamma rays of the radionuclides, in *Tables for Applied Gamma Ray Spectrometry*, Weinheim, New York, 1979.
4. Radioisotope Production and Quality Control, Technical Report Series No. 128, IAEA, Vienna, 1971.
5. **Ganiev, U., Artykbaev, T., and Tsyganov, G. A.**, Kinetics and products of the dissolution of molybdenum and tungsten in hydrogen peroxide, *Russian J. Inorganic Chem.*, 18(3), 370, 1973.
6. **Somerville, S. J.**, Mediphysics Inc., Emeryville, California, Private communication.
7. **Lewis, R. E. and Tenorio, J. I.**, Stable Solutions Containing Molybdenum-99 and Process of Preparing Same, US Patent 3,752,769 (14 August 1973).
8. **Strang, L. G., Jr.**, Manual of Isotope Production Processes in Use at Brookhaven National Laboratory, BNL-864 (T-347), Brookhaven National Laboratories, Upton, N.Y., 1964.
9. **Richards, P.**, The Technetium-99m Generator in *Radioactive Pharmaceuticals*, USAEC/Division of Technical Information, Oak Ridge, Tenn., 1966, 323.
10. **Ottinger, C. L.**, Short-Lived Fission Products Program, in CONF700646, Radioisotope Production Technology Development Meeting, Oak Ridge National Laboratories, Oak Ridge, Tenn., 1970.
11. **Kawakami, Y.**, Large scale production of ^{99}Mo, *Isotope News*, 6, 1977.
12. **Heyne, W.**, Spaltmolybdanabtrennung durch Extraktion mit Acetylaceton, *Isotopenpraxis*, 10(13), 343, 1977.
13. **Sivaramakrishnan, C. K., Jadhav, A. V., Raghuraman, K., Raman, S., Nair, P. S., and Ramanish, M. V.**, Preparation of High Purity Fission Produced Molybdenum-99, BARC-847, 1976.
14. **Tanase, M., Kase, T., and Shikata, E.**, Separation of molybdenum-99 from neutron-irradiated uranium-235 with sulfur as collector, *J. Nucl. Sci. Tech.*, 13(10), 591, 1976.
15. **Arino, H. and Kramer, H. H.**, Separation and purification of radiomolybdenum from a fission product mixture using silver-coated carbon granules, *Int. J. Appl. Radiat. Isotopes*, 29, 97, 1978.
16. **Barnes, R. K.**, personal communication of unpublished data.
17. **Tucker, D., Greene, M. W., Weiss, A. J., and Murrenhoff, A.**, Methods of Preparation of Some Carrier-Free Radioisotopes Involving Sorption on Alumina, BNL-3746, Brookhaven National Laboratories, Upton, N.Y., 1958.
18. **Boyd, R. E.**, Recent developments in generators of ^{99m}Tc, in *Radiopharmaceuticals and Labelled Compounds*, Vol. 1, Proc. Symp. New Developments in Radiopharmaceuticals and Labelled Compounds, IAEA, Vienna, 1973.

19. **Boyd, R. E. and Hetherington, E. L. R.**, Technetium-99m generators — operational assessment by performance indices, *Int. J. Appl. Radiat. Isotopes,* 31, 250, 1980.
20. **Vesely, P. and Cifka, J.**, Some chemical and analytical problems connected with technetium-99m generators, in *Radiopharmaceuticals from Generator-Produced Radionuclides,* Proc. Panel Preparation and Control of Radiopharmaceuticals from Generator-Produced Radionuclides, IAEA, Vienna, 1971.
21. **O'Donnell, J. H. and Sangster, D. F.**, *Principles of Radiation Chemistry,* Edward Arnold, London, 1970.
22. **Matthews, R. W.**, personal communication.
23. **Rhodes, B. A. and Croft, B. Y.**, *Basics of Radiopharmacy,* C.V. Mosby Co., St. Louis, 1978, 120.
24. **Boyd, R. E. and Matthews, R. W.**, Performance Aspects of Technetium-99m generators, in *Radiopharmaceuticals,* Australian National University, Canberra, 1977, 98.
25. **Panek-Finda, H.**, Isotope Generator Provided with a Carrier Material which in Addition to Al_2O_3 Contains Fully or Partly Hydrated MnO_2, US Patent 3,970,583 (20 July 1976).
26. **Boyd, R. E. and Matthews, R. W.**, Technetium-99m Generators — Improvements to Performance, US Patent 4,206,358 (16 October 1978).
27. **Meinhold, H., Herzberg, B., Kaul, A., and Roedler, H. D.**, Radioactive impurities of nuclide generators and estimation of resulting absorbed dose in man, in *Radiopharmaceuticals and Labelled Compounds,* Vol. 1, Proc. Symp. New Developments in Radiopharmaceuticals and Labelled Compounds, IAEA, Vienna, 1973.
28. **Finck, R. and Mattson, S.**, Long-lived radionuclide impurities in eluates from molybdenum-technetium generators and the associated absorbed dose, to the patient, *Int. J. Nucl. Med. Biol.,* 3, 89, 1976.
29. **Wood, D. E. and Bowen, B. M.**, ^{95}Sr and ^{124}Sb in ^{99}Mo-^{99m}Tc generators, *J. Nucl. Med.,* 12(6), 307, 1971.
30. **Billinghurst, M. W. and Hreczuch, F. W.**, Contamination from ^{131}I, ^{103}Ru and ^{239}Np in the eluate of ^{99}Mo^{99m}Tc generators loaded with (n,γ)-produced ^{99}Mo, *J. Nucl. Med.,* 17(9), 840, 1976.
31. **Hetherington, E. L. R. and Wood, N. R.**, SPECT, A FORTRAN Program for the Analysis of ^{99m}Tc and Other Gamma Spectra in a Radioisotope Quality Control Environment, AAEC/E319, 1974.
32. **Sodd, V. J. and Fortman, D. L.**, Analysis of the ^{89}Sr and ^{90}Sr Content in Eluates of Fission Produced ^{99}Mo-^{99m}Tc Generators, *Health Phys.,* 30, 179, 1976.
33. **Velten, R. J.**, Resolution of ^{89}Sr and ^{90}Sr in environmental media by an instrumental technique, *Nucl. Instr. Meth.,* 42, 169, 1966.
34. **Sodd, V. J. and Fortman, D. L.**, An investigation of the ^{89}Sr and ^{90}Sr contamination in fission-product ^{90}Mo-^{99m}Tc generators, *J. Nucl. Med.,* 16(6), 571, 1975.
35. **Charlton, J. C.**, Discussion, in *Radiopharmaceuticals and Labelled Compounds,* Vol. I, Proc. Symp. New Developments in Radiopharmaceuticals and Labelled Compounds, IAEA, Vienna, 1973, 51.
36. **Sodd, V. J., Grant, R. J., and Montgomery, D. M.**, An investigation of the plutonium content in fission product ^{99}Mo-^{99m}Tc generators, *Health Phys.,* 29, 425, 1975.
37. **Binney, S. E. and Scherpelz, R. I.**, A review of the delayed fission neutron technique, *Nucl. Instr. Meth.,* 154, 413, 1978.
38. **Shukla, S. K.**, Ion exchange paper chromatography of Tc(IV), Tc(V), and Tc(VII) in hydrochloric acid, *J. Chromatogr.,* 21, 92, 1966.
39. **Shen, V., Hetzel, K. R., and Ice, R. D.**, Radiochemical Purity of Radiopharmaceuticals Using Gelman Septrachrom (ITLC) Chromatography, Technical Bulletin 32, Gelman Instrument Company, Ann Arbor, Michigan, 1975.
40. **Weinstein, M. B. and Smoak, W. M., III**, Technical Difficulties in ^{99m}Tc-labelling of erythrocytes, *J. Nucl. Med.,* 111(1), 41, 1970.
41. **Lin, M. S., MacGregor, R. D., Jr., and Yano, Y.**, Erythrocyte agglutination by ionic Al(III) in generator eluate, *J. Nucl. Med.,* 12(6), 297, 1971.
42. **Samuels, L. D. and Hipple, T. H.**, A safe, rapid preparation method for ^{99m}Tc-sulfur colloid, *J. Nucl. Med.,* 11(4), 182, 1970.
43. **Haney, T. A., Ascanio, I., Gigliotti, J. A., Gusmano, E. A., and Bruno, G. A.**, Physical and biological properties of a ^{99m}Tc-sulfur colloid preparation containing disodium edetate, *J. Nucl. Med.,* 12(2), 64, 1971.
44. **Craigin, M. D., Webber, M. M., and Victery, W. K.**, Effect of aluminium concentration in technetium eluant on particle size, *J. Nucl. Med.,* 12(6), 476, 1971.
45. **Webber, M. M., Craigin, M. D., and Victery, W. K.**, Aluminium content in eluants from commercial technetium generators, *J. Nucl. Med.,* 12(10), 700, 1971.
46. **Chaudhuri, T. K.**, The effect of aluminium and pH on altered body distribution of ^{99m}Tc-EHDP, *Int. J. Nucl. Med. Biol.,* 3, 37, 1976.
47. **Chaudhuri, T. K.**, Liver uptake of ^{99m}Tc-diphosphonate, *Radiology,* 119, 485, 1976.
48. **Shukla, S. K., Mani, G. B., and Cipriani, C.**, Effect of aluminium impurities in the generator-produced pertechnetate-99m ion on thyroid scintigrams, *Eur. J. Nucl. Med.,* 2, 137, 1977.
49. **Turco, S. and King, R. E.**, *Sterile Dosage Forms, Their Preparation and Clinical Application,* Lea & Febiger, Philadelphia, 1979, 191.

50. **Charlton, J. C.**, Problems characteristic of radioactive pharmaceuticals, in *Radioactive Pharmaceuticals*, USAEC/Division of Technical Information, Oak Ridge, Tenn., 1966.
51. **Soresen, K., Kristensen, K., and Frandsen, P.**, Microbial contamination of radionuclide generators, *Eur. J. Nucl. Med.*, 2, 105, 1977.
52. **Saunders, M.**, AAEC — Personal communication of unpublished data.
53. **Morgan, F. and Sizeland, M. L.**, Tracer Experiments on Technetium, Report AERE C/M96, Atomic Energy Research Establishment, Harwell, U.K., 1950.
54. **Robson, J. and Boyd, R. E.**, The Production of Technetium-99m, in *Radioisotope Production*, IAEA, Vienna, 1969, 187.
55. **Robson, J.**, Process for the Production of Technetium-99m from Neutron Irradiated Molybdenum Trioxide, US Patent 3,833,469 (3 September 1974).
56. **Lee, E.**, Personal communication of unpublished data.
57. **Tachimori, S., Nakamura, H., and Amano, H.**, Diffusion of Tc-99m in neutron irradiated molybdenum trioxide and its application to separation, *J. Nucl. Sci. Technol.*, 8(6), 295, 1971.
58. **Vlcek, J., Rusek, V., Machan, V., Rohacek, J., Smejkal, Z., Kokta, L., and Vitkova, J.**, Thermal separation of ^{99m}Tc from molybdenum trioxide. I. Separation of ^{99m}Tc from molybdenum trioxide at temperatures below 650°C, *Radiochem. Radioanal. Lett.*, 20(1), 15, 1974.
59. **Vlcek, J., Machan, V., Rusek, V., Kokta, L., Rohacek, J., Smejkal, Z., and Vitkova, J.**, Thermal separation of ^{99m}Tc from molybdenum trioxide. II. Separation of ^{99m}Tc from molybdenum trioxide at temperatures above 650°C, *Radiochem. Radioanal. Lett.*, 20(1), 23, 1974.
60. **Machan, V., Vlcek, J., Kokta, L., Rusek, V., Smejkal, Z., Rohacek, J., and Vitkova, J.**, Thermal separation of ^{99m}Tc from molybdenum trioxide. III. Diffusion separation of ^{99m}Tc from molybdenum trioxide from the standpoint of its possible use in technetium generator, *Radiochem. Radioanal. Lett.*, 20(1), 33, 1974.
61. **Vlcek, J., Rusek, V., Vanickova, V., Vitkova, J., Smejkal, Z., Rohacek, J., Kokta, L., and Machan, V.**, Thermal separation of ^{99m}Tc from molybdenum trioxide. IV. Diffusion of ^{99m}Tc from molybdenum trioxide. Application for greater amounts of MoO$_3$, *Radiochem. Radioanal. Lett.*, 25(3), 173, 1976.
62. **Vlcek, J., Rusek, V., and Vanickova, V.**, Thermal separation of ^{99m}Tc from molybdenum trioxide. V. Thermal separation of ^{99m}Tc from molybdenum trioxide using a carrier-gas, *Radiochem. Radioanal. Lett.*, 25(3), 179, 1976.
63. **Colombetti, L. G., Husak, V., and Dvorak, V.**, Study of the purity of ^{99m}Tc sublimed from fission ^{99}Mo and the radiation dose from the impurities, *Int. J. Appl. Radiat. Isotopes*, 25, 35, 1974.
64. **Gerlit, J. B.**, Some chemical properties of technetium, *Proc. Int. Conf. Peaceful Uses of At. Energy*, 7, 145, 1956.
65. **Allen, J. F.**, An improved technetium-99m generator for medical applications, *Int. J. Appl. Radiat. Isotopes*, 16, 334, 1965.
66. **Harper, P. V., Lathrop, K. A., Jiminex, D., Fink, R., and Gottschalk, A.**, Technetium-99m as a scanning agent, *Radiology*, 85, 101, 1965.
67. **Anwar, M., Lathrop, K., Ross-Kelly, D., and Harper, P. V.**, Pertechnetate production from ^{99}Mo by liquid-liquid extraction, *J. Nucl. Med.*, 9(6), 298, 1968.
68. **Crews, M. C., Westerman, B. R., and Quinn, J. L., III**, Solvent extraction of ^{99m}Tc in the clinical laboratory, *J. Nucl. Med.*, 11(6), 386, 1970.
69. **Lathrop, K. A.**, Preparation and control of ^{99m}Tc radiopharmaceuticals, in *Radiopharmaceuticals from Generator-Produced Radionuclides*, Proc. Panel on Preparation and Control of Radiopharmaceuticals from Generator-Produced Radionuclides, IAEA, Vienna, 1971.
70. **Robinson, G. D.**, Simple manual system for the efficient routine production of ^{99m}Tc by methyl-ethyl-ketone extraction, *J. Nucl. Med.*, 12(6), 459, 1971.
71. **Baker, R. J.**, A system for the routine production of concentrated technetium-99m by solvent extraction of molybdenum-99, *Int. J. Appl. Radiat. Isotopes*, 22, 483, 1971.
72. **Tachimori, S., Amano, H., and Nakamura, H.**, Preparation of Tc-99m by direct adsorption from organic solution, *J. Nucl. Sci. Technol.*, 8(7), 357, 1971.
73. **Yang, J. Y., Tseng, C. L., and Yang, M. H.**, Separation of ^{99m}Tc from ^{99}Mo-^{99m}Tc mixture by solvent extraction with lubricating base oil, *Radiochem. Radioanal. Lett.*, 29(3), 111, 1977.
74. **Mikulaj, V., Macasek, F., and Steinerova, M.**, Chelate extraction in repeating separations of ^{99m}Tc from parent ^{99}Mo using N-benzoyl-N-phenylhydroxylamine, *Radiochem. Radioanal. Lett.*, 29(4), 199, 1977.
75. **Sanad, W., Tadros, N., and Haggag, A.**, A simple technetium generator, *J. Radioanal. Chem.*, 50(1-2), 153, 1979.
76. **Toren, D. M. and Powell, M. R.**, Automatic production of ^{99m}Tc for pharmaceutical use, *J. Nucl. Med.*, 11(6), 368, 1970.
77. **Charlier, R., Fallais, C., and Constant, R.**, Appareil Automatique Pour l'Extraction Liquide-Liquide du Technetium 99m, Rapport IRE (March 1973).

78. **Sorby, P. J. and Boyd, R. E.**, The Production of Approved Radiopharmaceuticals and the Development of New Radiopharmaceuticals, in Radiopharmaceuticals, Proceedings of a Seminar held at the Australian National University, Canberra, 1977.
79. **Hunter, W. W., Jr.**, Discussion, in *Radiopharmaceuticals and Labelled Compounds*, Proc. Symp. New Developments in Radiopharmaceuticals and Labelled Compounds, IAEA, Vienna, 1973.
80. **Narasimhan, D. V. S. and Mani, R. S.**, Chemical and radiochemical evaluation of the purity of ^{99m}Tc extracted by MEK, *J. Radioanal. Chem.*, 33, 81, 1976.
81. **Yeates, D. B., Warbick, A., and Aspin, N.**, Production of ^{99m}Tc labelled albumin microspheres for lung clearance studies and inhalation scanning, *Int. J. Appl. Radiat. Isotopes*, 25, 578, 1974.
82. **Wieland, H. C., Grames, G. M., Jansen, C., and Davidson, T.**, An efficient method for fractional labelling of microspheres, *J. Nucl. Med.*, 15(9), 808, 1974.
83. **Evans, J. V. and Matthews, R. W.**, Technetium-99m Generator, US Patent Application No. 912,146.
84. **Richards, P. and Steigman, J.**, *Chemistry of Technetium as Applied to Radiopharmaceuticals*, Subramanian, G., Rhodes, B. A., Cooper, J. F., and Sodd, V. J., Eds., Society of Nuclear Medicine, New York, 1975, 23.
85. **Eckelman, W. C. and Levenson, S. M.**, Radiopharmaceuticals labelled with technetium, *Int. J. Appl. Radiat. Isotopes*, 28, 67, 1977.
86. **Vilcek, S., Machan, V., Kalincak, M., and Nicak, A.**, ^{99m}Tc-Ti-DTPA: preparation, control and biological distribution, *Int. J. Appl. Radiat. Isotopes*, 30, 673, 1979.
87. **Agha, N. H., Al-Hilli, A. M., and Hassen, H. A.**, A new technetium-99m-EDTA complex production technique for renal studies, *Int. J. Appl. Radiat. Isotopes*, 30, 353, 1979.
88. **Fritzberg, A. R., Lyster, D. M., and Dolphin, D. H.**, Evaluation of formamidine sulfinic acid and other reducing agents for use in the preparation of Tc-99m labelled radiopharmaceuticals, *J. Nucl. Med.*, 18(6), 553, 1977.
89. **Sundrehagen, E.**, Polymer formation and hydrolysation of ^{99}Tc(IV), *Int. J. Appl. Radiat. Isotopes*, 30, 739, 1979.
90. **Ekelman, W., Meinken, G., and Richards, P.**, Chemical state of ^{99m}Tc in biomedical products, *J. Nucl. Med.*, 12, 596, 1971.
91. **Srivastava, S. C., Meinken, G., Smith, T. D., and Richards, P.**, Problems associated with stannous ^{99m}Tc-radiopharmaceuticals, *Int. J. Appl. Radiat. Isotopes*, 28, 83, 1977.
92. **Owunwanne, A., Church, L. B., and Blau, M.**, Effect of oxygen on the reduction of pertechnetate by stannous ion, *J. Nucl. Med.*, 18, 822, 1977.
93. **Owunwanne, A., Marinsky, J., and Blau, M.**, Charge and nature of technetium species produced in the reduction of pertechnetate by stannous ion, *J. Nucl. Med.*, 18, 1099, 1977.
94. **Russell, C. D. and Cash, A. G.**, Oxidation State of Technetium in Bone Scanning Agents, in Proc. 2nd Int. Radiopharmaceutical Symp., Seattle, 1979.
95. **Deutsch, E.**, Inorganic Radiopharmaceuticals, in Proc. 2nd Int. Radiopharmaceutical Symp., Seattle, March 1979.
96. **Benjamin, P. P.**, A rapid and efficient method of preparing ^{99m}Tc-human serum albumin: its clinical applications, *Int. J. Appl. Radiat. Isotopes*, 20, 187, 1969.
97. **Benjamin, P. P., Rejali, A., and Friedell, H.**, Electrolytic complexation of ^{99m}Tc at constant current: its applications in nuclear medicine, *J. Nucl. Med.*, 11, 147, 1970.
98. **Steigman, J., Eckelman, W. C., Meinken, G., Isaacs, H. S., and Richards, P.**, The chemistry of technetium labelling of radiopharmaceuticals by electrolysis, *J. Nucl. Med.*, 15, 75, 1974.
99. **Scheider, P. B.**, A simple electrolytic preparation of a ^{99m}Tc(Su)-citrate renal scanning agent, *J. Nucl. Med.*, 14, 843, 1973.
100. **Narasimhan, D. V. S. and Mani, R. S.**, Electrolytic preparation of ^{99m}Tc human serum albumin using tin electrodes, *Radiochem. Radioanal. Lett.*, 26, 307, 1975.
101. **Gil, M. C., Palma, T., and Radicella, R.**, Electrolytical labelling of ^{99m}Tc radiopharmaceuticals, *Int. J. Appl. Radiat. Isotopes*, 27, 69, 1976.
102. **Russell, C. D. and Majerik, J.**, Tracer electrochemistry of pertechnetate: chelation of ^{99m}Tc by EDTA after controlled-potential reduction at mercury and platinum cathodes, *Int. J. Appl. Radiat. Isotopes*, 29, 109, 1978.

Chapter 5

PRODUCTION OF RADIONUCLIDES BY 14 MeV NEUTRON GENERATOR

Zeev B. Alfassi

TABLE OF CONTENTS

I.	Introduction	154
II.	14 MeV Neutron Generators	154
III.	Radionuclide Production	155
IV.	Separation Methods	156
V.	Preparation Processes	158
	A. Production of ^{18}F	158
	B. Production of ^{34m}Cl	158
	C. Production of Radiobromine Nuclides	158
	D. Production of ^{128}I	159
	E. Production of ^{28}Al	159
	F. Production of ^{27}Mg	159
	G. Production of ^{56}Mn	159
VI.	Summary	160
Acknowledgment		160
References		160

I. INTRODUCTION

14 MeV neutron generators are useful for preparation of short-lived radionuclides ($t_{1/2} <$ ~12 hr) only. This is due to their advantage of simplicity of operation and relatively low cost and disadvantage of relatively low yield.

Short-lived radionuclides are widely used for clinical work due to their virtue of lower dose delivered to the patient. Besides, for some elements the use of short-lived radionuclides is a must due to lack of long-lived ones (nitrogen, oxygen, fluorine) or very long-lived ones such as $^{26}Al(7.4 \cdot 10^5 y)$ or $^{36}Cl(3.1 \cdot 10^5 y)$.[1] For biological and chemical studies short-lived radionuclides also have the advantage of less hazard of radioactive contamination.

Due to the short half-lives of these nuclides they have to be produced *in situ* or at least not far from the place of use. The cost of 14 MeV neutron generators have been compared[2] with the typical middle-sized cyclotrons and it was found that the capital costs are much lower in the case of neutron generators. This is the main reason for the availability of 14 MeV neutron generators in many scientific institutes compared to the scarcity of cyclotrons. Lately, the use of 14 MeV neutrons for cancer therapy was studied in several medical centers. A number of hospitals and cancer research centers have high intensity 14 MeV neutron generators for this purpose. The advantages of using short-lived in-house produced radionuclides suggest the use of the available 14 MeV neutron generators for biological studies and in medical diagnosis.

The main disadvantage of 14 MeV neutron generators is their low flux. Small neutron generators produce 10^{10}-10^{11} n·sec^{-1} which leads to fluxes of 10^8 n·sec^{-1}·cm^{-2}, while new high output 14 MeV neutron generators yield ~10^{13} n·sec^{-1} with fluxes of ~10^{11} n·sec^{-1}, well below the fluxes of nuclear reactors and cyclotrons. Besides, the cross-sections for the reactions of fast neutrons are considerably lower than those of thermal neutrons or charged particles. These disadvantages can be partly overcome by the irradiation of large samples (usually above 250 g) in order to obtain sufficient amounts of radioactivity. As a result, the irradiated material cannot be used without processing and the nuclide produced has to be separated from the target.[3]

II. 14 MeV NEUTRON GENERATORS

Generation of neutrons of approximately 14 MeV energy is based on the exothermic fusion reaction between deuterium and tritium to yield a neutron and an alpha-particle. The relative kinetic energy to overcome the Coulombic barrier is about 60 keV and typical generators operate in the range of 150 to 250 keV; the cross-section for the reaction peaks at about 110 keV (5 barns). The accelerating system for the deuteron beam includes an ion source, accelerating structure, vacuum pump to remove residual gas, a beam tube at ground potential, and a target isolation valve. The tritium target usually consists of a thin layer of material which absorbs hydrogen nuclides, such as titanium or zirconium, deposited on a material such as copper. The backing material prevents diffusion of the hydrogen nuclides and allows good heat conduction to the cooling system behind the target. The cooling is essential to remove the substantial amount of heat produced by the incident beam. With an accelerator which produces deuteron beams of 150 to 250 keV and a target with approximately a 1:1 atomic ratio of tritium to titanium the neutron yield is 10^{11} n/sec per mA of the beam.

Sealed-tube neutron generators overcome some of the difficulties of short target life-time, the need for continuous pumping, and the need for tritium contamination control equipment. The tube is filled with a mixture of deuterium and tritium gas so that the beam and the target both contain approximately equal amounts of deuterium and tritium. Instead of a pump, a gas reservoir is included, usually of titanium, which can be heated to bring the tube up to the operating pressure, or cooled off to soak up residual gas in the tube.

The energy of the neutrons depends on the energy of the incident deuterons and on the angle between the beam of the deuteron and the direction of the emitted neutrons. Usual work uses angles of 0 to 70° and the energy of the neutron ranges between 15.2 to 14.3 MeV.

III. RADIONUCLIDE PRODUCTION

Due to the high energy of the neutrons, the (n,α) (n,p) and $(n,2n)$ reactions are possible. For low Z elements the $(n,2n)$ reactions have a cross-section which can be as low as a few mb and the (n,p) and (n,α) reactions are the predominant ones.[4] For medium and high Z elements the $(n,2n)$ reactions have cross-sections of several hundred mb compared to usually few mb for (n,p) and (n,α) reactions. Thus the main source of radionuclides is the $(n,\alpha n)$ reactions and only a few contributions from (n,p) and (n,α). The $(n,2n)$ reaction produces neutron-deficient nuclides which decay by positron emission and thus are candidates for use in positron tomography.

Due to the relatively low fluxes and cross-sections, large samples have to be irradiated. The fast neutrons are slowed-down in these large quantities of material and we cannot speak of the same flux over all the target material, as it is done in nuclear reactors. Usually an effective flux is used, but it must be remembered that this effective flux depends on the size of the sample as well as its chemical composition. The larger the fraction of hydrogen atoms, the higher is the moderation and thus a lower flux of high energy neutrons is obtained; for such high energy inelastic scattering is also important[5] and the moderation by heavy atoms must also be taken into consideration. For average size samples, 500 to 1000 g — in cylinders of 4 to 5 cm radius and 4 to 5 cm thick — the effective flux is about $3 \cdot 10^{-2}$ to 10^{-3} of the rate of production of 14 MeV neutrons. High intensity neutron generators produce 10^{13} $n \cdot sec^{-1}$ with effective flux of about $2 \cdot 10^{10}$ $n \cdot sec^{-1} \cdot cm^{-2}$. Tables 1 to 3 give lists of possible short-lived radio-nuclides (2 min $< t_{1/2} <$ 12 hr) together with their nuclear emission and activities obtained in irradiation of 1 hr or two half-lives, the shorter of the two. Longer irradiations than 1 hr are not usually done due to the limited life of the sealed tube or the target (tube life = 100 to 200 hr, target life — 20 to 50 hr). Most of the radionuclides are produced by $(n,2n)$ reaction and have a high percentage of β^+.

Besides the more common reactions (n,p), (n,α) and $(n,2n)$, some radionuclides can be produced by the (n,n') and the (n,γ) reactions. Some of the inelastic scattering reactions (n,n') are used frequently in activation analysis by 14 MeV neutrons and found to offer good analytical sensitivity,[6] indicating considerable activity. Interesting radionuclides with reasonable half-lives which can be produced by this reaction are ^{199m}Hg (42 min, main γ lines 0.58 MeV (52%) and 0.374 MeV (15%)[1]) and ^{204m}Pb (67 min, main γ lines 0.379 MeV (93%) 0.899 MeV (99%) and 0.912 MeV (97%)). These radionuclides have the virtue of decaying only by isomeric transition and hence delivering lower dose to the patient than β^- or β^+ emitters. Besides they have low γ-rays, mainly ^{199m}Hg, which are well fitted for usual γ-ray cameras.

The (n,γ) cross-sections are very low for such high energy neutrons and higher activity will be obtained if the neutrons are thermalized by paraffins before hitting the target. This thermalization leads to a flux which is about 10^{-4} of the rate of production of the neutrons, i.e., about 10^9 $n \cdot sec^{-1} \cdot cm^{-2}$ for high intensity generators. For some elements high energy reactions do not lead to a desirable half-life, while the (n,γ) reaction does so. An example is iodine where $(n,2n)$ reaction leads to ^{126}I with a half-life of 13 days compared to the 25 min ^{128}I which is produced by the (n,γ) reaction. However, these are the same radionuclides produced by nuclear reactors and it will not be treated more in this chapter.

Some other reactions are also possible such as $(n,^3He)$ and (n,t),[7] however, these reactions have very low cross-sections of several μb and hence cannot be used for production of radionuclides with reasonable activities.

Table 1
SHORT-LIVED RADIONUCLIDES PRODUCED BY (n,p) REACTION WITH 14 MeV NEUTRONS

Radionuclide	Target	Isotopic abundance (%)	$T_{1/2}$	Cross section (mb)	β^- Energy (MeV)	Main γ lines (keV)	Yield[a] (mCi/100 gr)
^{27}Mg	^{27}Al	100	9.46 min	78	1.75	840(70%),1013(30%)	70.2
^{28}Al	^{28}Si	92.2	2.3 min	250	2.85	1780(100%)	200.0
^{29}Al	^{29}Si	4.7	6.6 min	120	2.40	1280 (94%)	4.7
^{31}Si	^{31}P	100	2.62 hr	88	1.48	1260 (0.7%)	21.5
^{37}S	^{37}Cl	24.5	5.05 min	25	1.6	3090 (90%)	4.0
^{41}Ar	^{41}K	6.9	1.83 hr	50	1.20	1293 (99%)	0.9
^{51}Ti	^{51}V	99.8	5.8 min	35	2.14	320 (95%)	16.7
^{52}V	^{52}Cr	83.8	3.75 min	90	2.47	1434(100%)	35.3
^{55}Cr	^{55}Mn	100	3.6 min	45	2.59	No γ	19.9
^{56}Mn	^{56}Fe	91.7	2.57 hr	112	2.85	847(99%),1811(29%)	14.1
^{58M}Co	^{58}Ni	67.9	9.0 hr	218	no β^-	Co X-rays	6.0
^{60M}Co	^{60}Ni	26.2	10.5 m	25	1.55(0.25%)	59 keV (2.1%)	2.7
^{65}Ni	^{65}Cu	30.9	2.56 hr	24	2.13	1481(25%),1115(16%)	0.9
^{66}Cu	^{66}Zn	27.8	5.1 min	75	2.63	1039(9%)	7.7
^{69}Zn	^{69}Ga	60.4	57 min	17	0.90	No γ	2.5
^{70}Ga	^{70}Ge	20.5	21.1 min	110	1.65	1040(0.5%)	7.8
^{75}Ge	^{75}As	100	83 min	21	1.19	265(11%)	3.6
^{88}Rb	^{88}Si	82.6	17.8 min	17	5.3	1863(21%),898(2.5%)	3.9

[a] Calculated for flux of $2 \cdot 10^{10}$ n·sec^{-1}·cm^{-2} and irradiation time of 1 hr or two half-lives, the shorter among the two.

Table 2
SHORT-LIVED RADIONUCLIDES PRODUCED BY (n,α) REACTION WITH 14 MeV NEUTRONS

Radionuclide	Target	Isotopic abundance (%)	$T_{1/2}$	Cross section (mb)	β^- Energy (MeV)	Main γ lines (keV)	Yield[a] (mCi/100 gr)
^{27}Mg	^{30}Si	3.1	9.46 min	70	1.75	840(70%),1013(30%)	1.8
^{28}Al	^{31}P	100	2.3 min	119	2.85	1780(100%)	93.3
^{31}Si	^{34}S	4.2	2.62 hr	126	1.48	1260(0.7%)	1.2
^{38}Cl	^{41}K	6.9	37.2 min	46	4.91	1600(38%),2170(47%)	1.7
^{52}V	^{55}Mn	100	3.75 min	32	2.47	1434(100%)	14.1
^{56}Mn	^{59}Co	100	2.57 hr	30	2.85	847(99%),1811(29%)	4.5
^{137M}Ba	^{140}Ce	88.5	2.55 min	12	no β	662(89%)	1.8

[a] Calculated for flux of $2 \cdot 10^{10}$ n·sec^{-1}·cm^2 and irradiation time of the shorter between two half-lives and 1 hr.

IV. SEPARATION METHODS

Two factors have to be considered when choosing methods for separation of the newly produced isotopes. These factors are the large sample irradiated; usually about 500 g, and the decay of the short-lived radionuclides which demands rapid separation processes. Appropriate processes are elution from a solid target on a column by an appropriate solution or solvent extraction for liquid targets. Figure 1 shows the vessels used for these methods. It is preferable to use polyethylene vessels since silica containing materials become highly activated due to ^{28}Al, ^{29}Al, and ^{27}Mg, or to use remote electrical operation.

Table 3
SHORT-LIVED RADIONUCLIDES PRODUCED BY (n,2n) REACTION WITH 14 MeV NEUTRONS

Radionuclide	Target	Isotopic abundance (%)	$T_{1/2}$	Cross section (mb)	β Emission abundance (%) and energy (MeV)	Main γ lines (keV)	Yield[a] (mCi/100 gr)
^{13}N	^{14}N	100	9.96 min	6.1	100% β$^+$, 1.20	511 (200%)	10.5
^{18}F	^{19}F	100	109.7 min	47	97% β$^+$, 0.635	511 (194%)	25.3
^{30}P	^{31}P	100	2.50 min	10	100% β$^+$, 3.24	511(200%),2.23(0.5%)	7.8
^{34m}Cl	^{35}Cl	75.5	32.0 min	4	53% β$^+$, 2.48n	511(106%),145(45%), 2120(38%)	2.1
^{62}Cu	^{63}Cu	69.1	9.76 min	480	97.8% β$^+$, 2.93	511(196%),880(0.3%)	128
^{63}Zn	^{64}Zn	48.9	38.4 min	119	93% β$^+$, 2.34	511(186%),669(8%), 962(6%)	19.5
^{68}Ga	^{69}Ga	60.4	68.3 min	850	88% β$^+$, 1.90	511(176%),1078(3.5%)	111.6
^{75}Ge	^{76}Ge	7.8	83 min	1157	β$^-$, 1.19	265(11%),199(1.4%)	15.2
^{81}Se	^{82}Se	9.2	18.4 min	225	β$^-$, 1.58	280 (0.9%)	6.1
^{81m}Se	^{82}Se	9.2	57 min	894	No β	103 (8%)	24.4
^{78}Br	^{79}Br	50.5	6.4 min	862	92% β$^+$, 2.55	511(184%),614(14%)	134.0
^{80}Br	^{81}Br	49.5	17.6 min	390	92% β$^-$ + 2.6% β$^+$	511 (5.2%),618 (7%)	58.0
^{84m}Rb	^{85}Rb	72.1	21 min	400	No β	250(65%),216(37%)	82.5
^{95}Ru	^{96}Ru	5.5	1.65 hr	774	15% β$^+$, 1.33	340(70%), 511(30%),1090(21%)	15.2
^{106}Ag	^{107}Ag	51.4	24 min	900	70% β$^+$, 1.96	511 (140%)	105.0
^{108}Ag	^{109}Ag	48.7	2.4 min	740	97.5% β$^-$ + 0.28% β$^+$	632(1.7%)	80.4
^{111m}Cd	^{112}Cd	24.1	49 min	624	No β	247(94%),150(30%)	24.9
^{129}Te	^{130}Te	34.5	69 min	600	β$^-$, 1.60	690 (6%)	23.4
^{135m}Xe	^{136}Xe	8.87	15.6 min	750	No β	527 (80%)	11.9
^{137m}Ba	^{138}Ba	71.7	2.55 min	1020	No β	662 (89%)	129
^{140}Pr	^{141}Pr	100	3.4 min	1800	50% β$^+$, 2.32	511(100%), 1596(0.3%)	310
^{197m}Pt	^{198}Pt	7.2	94.5 min	1080	3% β$^-$	346 (13%)	4.5
^{196m}Au	^{197}Au	100	9.7 h	134	No β	148 (42%), 188(32%)	1.5
^{199m}Hg[b]	^{200}Hg	23.1	43 min	880	No β	158 (53%),375 (15%)	20.4

[a] Calculated for flux of $2 \cdot 10^{10}$ n·sec^{-1}·cm^{-2} and irradiation time of 1 hr or two half-lives, the shorter of the two.
[b] Including also the ^{199}Hg (n,n') ^{199m}Hg reaction.

(n,p) and (n,α) reactions lead to new elements and chemical separation can be found relatively easily, usually by elution of an appropriate solid target material. (n,2n) and (n,n') lead to the same element and Szillard-Chalmers processes are essential. In these processes the atom undergoing a nuclear transformation breaks the bond to the residual part of the molecule due to its high recoil energy and forms an atomic or molecular species which is different and separable from the parent form. Many examples which have been studied for the (n,γ) reaction with nuclear reactors[8] can be used, except that with nitrogen containing ligands, the ^{14}N (n,2n)^{13}N reactions have to be taken care of.

The efficiency of the separation methods was determined by the measurements of the activities of the separated radionuclides and the total irradiated target, far enough from the detector such that geometrical corrections were negligible.

V. PREPARATION PROCESSES

All the preparation processes were developed with a small neutron generator of up to 10^{11} n/sec. The yields given are the ones obtained for this generator. High intensity neutron generators used for patient treatments will produce activities of 2 to 3 orders of magnitude higher.

The radiochemical purity was checked by γ-ray spectrometry and with standard chemical analysis.

A. Production of ^{18}F

370 g of perfluorohexane with, on top of them, 15 to 25 mℓ of aqueous solution of 0.01 M NaOH in a cylindrical vessel (4.3 cm radius and 4.2 cm thickness) were bombarded with 14 MeV neutrons (total output 5·10^{10} n/sec) for 30 min. After irradiation, the two phases in the vessel were mixed thoroughly and then separated. The aqueous phase (the upper one) was boiled down to about 5 to 10 mℓ to concentrate the ^{18}F and mainly to remove any traces of the organic compound. The concentrated aqueous solution has activity of about 20 μCi.[9]

Radiochemical purity — γ-ray spectrometry shows only the 511 keV line due to β$^+$ anihilation. Decay measurements give half-life of 108.7 min in agreement with previously reported values. Gas liquid chromatography does not show any perfluoro-n-hexane (less than 10^{-5} M). All the activity is precipitated by solution of $CaCl_2$. Thin-layer chromatography (cellulose adsorbent and 0.15 M sodium acetate as solvent[10]) shows only one peak of activity. The F$^-$ concentration was found to be about 10^{-4} M.

B. Production of ^{34m}Cl

320 g of carbon tetrachloride together with 0.1 g I_2 and with 20 to 25 mℓ of 0.01 M aqueous solution of NaOH were bombarded with 14 MeV neutrons (5·10^{10} n/sec) for 30 min. After irradiation the I_2 is reduced by titrating with 0.1 N solution of Na_2SO_3 till the color of I_2 disappears. The two phases were shaken vigorously and the two phases were separated. The aqueous phase was evaporated to 5 mℓ. The concentrated aqueous solution has activity of about 0.5 to 1.0 μCi^{34m}Cl. The separation procedure takes about 20 min.

Radiochemical purity — The γ-ray spectra shows the presence of ^{34m}Cl, together with ^{37}S produced by ^{37}Cl(n,p) ^{37}S, and ^{38}Cl due to ^{37}Cl(n,γ) ^{38}Cl. The cross-section for production of ^{37}S is higher than for formation of ^{34m}Cl, although its inorganic yield is lower. Due to its shorter half-life of 5.1 min, after delay of about 40 min from the end of the irradiation ^{37}S activity is less than 1% of the ^{34m}Cl activity. The activity of ^{38}Cl was found to be 1 to 2% of the ^{34m}Cl activity. If chloroform is used instead of CCl_4 as a target material the activity of ^{38}Cl is almost doubled. Cl$^-$ concentration was found to be 2·10^{-3} M.

C. Production of Radiobromine Nuclides

For ^{78}Br preparation, 220 cc of bromoethane with 12 cc of 0.01 N NaOH are irradiated for 10 min, shaken, separated, and boiled to about 10 cc. The solution contains about 40 μCi of 6.4 min ^{78}Br, 20 μCi of 17.6 min ^{80}Br, 7 μCi of ^{80m}Br, and about 20 μmole of Br. The separation procedure takes about 3 to 4 min.

For ^{80}Br and ^{80m}Br preparation, 0.1 g I$_2$ is added to the bromoethane in order to increase the inorganic yield. After irradiation of 30 min the I$_2$ is reduced by titration with 0.1 N thiosulfate solution. To increase the inorganic yield the extraction was done with 20 mℓ which were boiled down to 5 cc. The preparation takes about 20 min. The aqueous solution contains about 10 μCi of ^{78}Br, 150 μCi of ^{80}Br and 20 μCi of ^{80m}Br. Higher activities are obtained with bromoform (CHBr$_3$) and dibromomethane (CH$_2$Br$_2$) as target material. However, due to their high boiling point, their traces cannot be removed by partial evaporation.

D. Production of ^{128}I

This radionuclide is produced by (n,γ) reaction after thermalization of the neutrons. 250 cc of methyliodide with 0.1 g I$_2$ together with 15 cc of 0.01 N NaOH are irradiated behind a paraffin sheet of 3 cm thickness for 20 min. After irradiation the I$_2$ is reduced by titration with 0.1 N sodium sulfite solution and the solvents are separated. The aqueous phase is evaporated to 5 cc. The whole procedure takes about 20 min. The aqueous solution contains about 30 μCi of ^{128}I, 0.3 μCi of ^{126}I and 35 μmole of Γ.

E. Production of ^{28}Al

250 g of SiO$_2$ in a cylinder fitted with a sintered glass frit (Figure 1) is treated with 20 cc of 0.1 N HCl and most of the solution is pumped out. This target is irradiated for 4 min (^{28}Al) or 12 min (^{29}Al). After irradiation 10 cc of 0.1 N HCl is added to the cylinder to elute the newly produced radioaluminum. The elution is carried out under vacuum and the whole separation procedure takes less than 1 min. The short irradiation gives about 200 to 300 μCi of ^{28}Al and about 3 to 4 μCi of ^{29}Al. The longer irradiation gives about the same amount of ^{28}Al but about twice as much ^{29}Al. The HCl solution can be replaced by other acidic solutions, but a strong acidic solution is essential and 0.1 N acetic acid was found to be insufficient. 0.01 N HCl gives slightly lower yields.[11]

Radiochemical purity — When the silica was washed well before irradiating, no traces of Si were found in the eluted solution. Gamma spectroscopy showed that the only contamination is from ^{27}Mg (about 0.5 μCi for short irradiation and 1.0 μCi for the longer one).

F. Production of ^{27}Mg

250 g of Al$_2$O$_3$ in a cylinder with sintered frit is wetted with 5 mℓ of pure water. After irradiation for 20 min, 10 mℓ of water or saline solution is added to elute the ^{27}Mg. The eluted solution contains 50 to 70 μCi of ^{27}Mg with contamination of about 1 to 2 μCi of ^{24}Na. The elution carried out under vacuum takes less than 1 min. Chemical analysis shows 10^{-4} M of Al ions in the solution.[12]

G. Production of ^{56}Mn

500 g of ferrous sulfide in a cylinder fitted with sintered frit and connected to a rotary pump were washed with at least 3 portions of 100 cc of 0.1 N NH$_4$OH solution. Then it was irradiated for 1 hr. After irradiation, 150 cc of 0.1 N NH$_4$OH solution was added and the sample was filtered slowly under vacuum. A solution of 10 cc 0.1 N NaOH was added to the filtrate which was then evaporated to about 10 cc. The whole procedure took about 30 min, mostly for the evaporation.[13]

Radiochemical purity — γ-ray spectrometry immediately after irradiation showed traces of ^{53}Fe (8.5 min) and ^{13}N (10 min). The ^{13}N was produced from traces of NH$_4$OH left after the washings and can be eliminated if two washings by water followed the washings by NH$_4$OH solution. One hour after the irradiation, no other peaks besides those of ^{56}Mn were found to an accuracy of 1% of ^{56}Mn activity with a Ge(Li) detector. Atomic absorption spectrometry showed that the solution contained 7 to 15 µg Fe/mℓ. This is the reason for the previous washings by NH_4OH which was found to decrease the concentration of the Fe in the solution. The concentration of the ammonium ions in the solution was found to be less than 10 ppm.

VI. SUMMARY

14 MeV neutron generators can be used to produce some of the medically useful radionuclides, such as ^{18}F, ^{80}Br, ^{199m}Hg, and others. However, the amount required for medicine can only be prepared by the new high intensity neutron generators, used for neutron therapy and not by the smaller ones, commonly used in university laboratories ($\sim 10^{11}$ n/sec). On the other hand, these relatively small neutron generators can be used for the preparation of radionuclides for biological studies. They facilitate the study of metabolism of elements for which radionuclides cannot be usually purchased due to short half-lives or the high price of the long-lived ones, such as ^{34m}Cl, ^{18}F, $^{28,29}Al$, ^{27}Mg, and others. An example is the work done on the fate of Al and Mg in rats using ^{28}Al and ^{27}Mg.[13]

ACKNOWLEDGMENT

The author wants to thank the U.S.-Israel Binational Science Foundation for the grant which enabled this study.

REFERENCES

1. **Lederer, C. M. and Shirley, V. S.,** *Table of Isotopes,* 7th ed., John Wiley & Sons, New York, 1978.
2. **Kushelevsky, A. P., Alfassi, Z. B., Schlesinger, T., and Wolf, W.,** *Int. J. Appl. Radiat. Isotopes,* 30, 275, 1979.
3. **Alfassi, Z. B.,** *Proc. 7th Int. Conf. on Cyclotrons and their Applications,* Birkhauser, Basel, 1975, 465.
4. **Bormann, M., Neuert, H., and Scobel, W.,** *Handbook on Nuclear Activation Cross-Sections,* Technical Report Series No. 156, IAEA, Vienna, 1974, 87.
5. **Lamarsh, J. R.,** *Introduction to Nuclear Engineering,* Addison-Wesley, Reading, 1975, 50.
6. **Nargolwalla, S. S. and Przybylowicz, E. P.,** *Activation Analysis with Neutron Generators,* John Wiley & Sons, New York, 1973, 642.
7. **Qaim, S. H.,** *J. Inorg. Nucl. Chem.,* 36, 239, 1974; Qaim, S. H., Wölfle, R., and Stöcklin, G., *J. Inorg. Nucl. Chem.,* 36, 3639, 1974.
8. **Harbottle, G. and Hillman, M.,** *Radioisotope Production and Quality Control,* IAEA Technical Reports Series No. 128, IAEA, Vienna, 1971, 617.
9. **Alfassi, Z. B. and Kushelevsky, A. P.,** in press.
10. **Noto, M. P. and Nicolini, J. O.,** *J. Radioanal. Chem.,* 24, 85, 1975.
11. **Alfassi, Z. B. and Kushelevsky, A. P.,** *Radiochem. Radioanal. Lett.,* 20, 347, 1975.
12. **Alfassi, Z. B. and Kushelevsky, A. P.,** *Radiochem. Radioanal. Lett.,* 21, 87, 1975.
13. **Alfassi, Z. B.,** *Radiochem. Radioanal. Lett.,* 32, 321, 1978.
14. **Kushelevsky, A., Yagil, R., Alfassi, Z., and Berlyne, G. M.,** *Biomedicine,* 25, 59, 1976.

Chapter 6

RADIONUCLIDES AND LABELED COMPOUNDS PRODUCED AT AN ELECTRON LINEAR ACCELERATOR

E. L. Sattler

The predominant or most efficient nuclear reaction at an electron linear accelerator is the giant resonance of the bremsstrahlung, the (γ,n)-reaction. In addition, some other reactions may be useful (Table 1). The maximum of the giant resonance lies at above 20 MeV. A significant yield of this nuclear reaction requires 25 MeV or more electrons at 100 kW. The most evident drawback of the electron linear accelerator compared with other particle accelerators is the radioisotope production instead of radionuclide production. That is, the (γ,n)-reaction produces activity of the same chemical element, which by necessity will be of low specific activity. The latter being useful only in biology and medicine, namely with the biologically predominating elements like oxygen, chlorine, and potassium, for example, and the compounds like sugar and fats.

In order to isolate the radioisotopes (from the stable element) there exists one possibility originally demonstrated by Szillard and Chalmers. The nuclear reaction imparts recoil energy to the newly formed radioactive atom, thereby completely liberating it from its initial position and bonding. Using appropriate measures one will be able to get the radioactive species isolated. The recoiling atoms are in a very reactive form (hot atom chemistry) and will form new bonds or new ions, respectively.

This chapter will deal mainly with the formation of new molecules which will become radioactively labeled by the uptake of or reaction with the hot atom. In former times this method has been used mostly with the reaction of ^{14}C produced from nitrogen with fast neutrons by the (n,p)-reaction.[1,2] Only a few experiments have been performed using the (γ,n)-reaction.[3] Due to the low yield, and specific yield, the γ,n-reaction has become now more interesting since high pressure liquid chromatography has enabled the separation of gram amounts. Some specific activations do not need chromatographic separation. Thus, if the transformation involves the passage from the solid to the gaseous phase the separation will be very simple.

Carbonoxide can be produced by flushing with oxygen charcoal under irradiation.[4] Carbonmonoxide is primarily generated under this condition. This is easy to convert to carbon dioxide. Carbon dioxide is the substrate for the photosynthetic cycle. Sugars are quickly formed by photosynthesis in green plants. It is therefore possible to produce rapidly hexoses and saccharose via biosynthesis.

Granulated alumina (Al_2O_3) as target releases ^{15}O-atoms, which form labeled oxygen molecules within a surpassing oxygen stream, useful for metabolic studies simply by inhalation. The produced amounts of ^{11}C and ^{15}O depend on the amount and mass to surface ratio of the target material. Ten and more mCi/liter and minute should be achievable.

Irradiation of amino acids[5-7] creates a lot of activated substances. Depending on the analytical procedure one can distinguish up to eight main products. As much as 50% of the products behave electrophoretically like amino acids (Figure 1). One part (5 to 8%) of this portion consists of activated original amino acid (retention), which may have been labeled via the "billiard ball process" described by Libby.[8] Further 5 to 8% could be identified as chain elongation products, probably via insertion (Figure 2). These investigations (of our group at Giessen) are at the moment restricted to aliphatic neutral amino acids. The remaining products have not yet been identified.

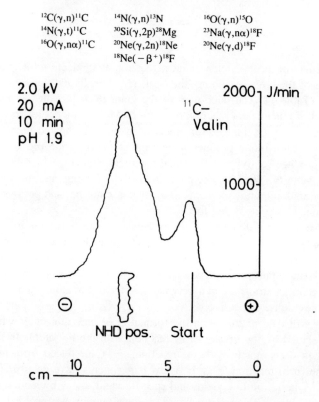

Table 1
SOME NUCLEAR REACTIONS OF BIOMEDICAL INTEREST AT AN ELECTRON LINEAR ACCELERATOR

$^{12}C(\gamma,n)^{11}C$ $^{14}N(\gamma,n)^{13}N$ $^{16}O(\gamma,n)^{15}O$
$^{14}N(\gamma,t)^{11}C$ $^{30}Si(\gamma,2p)^{28}Mg$ $^{23}Na(\gamma,n\alpha)^{18}F$
$^{16}O(\gamma,n\alpha)^{11}C$ $^{20}Ne(\gamma,2n)^{18}Ne$ $^{20}Ne(\gamma,d)^{18}F$
$^{18}Ne(-\beta^+)^{18}F$

FIGURE 1. Activation of the amino acid valine. High voltage paper electrophoresis showing two activity peaks: one at the starting position and one migrating like valine and other neutral amino acids.

The same type of reactions could be found by treating short chained amines. Here again some percent are retention, that is $^{11}C/^{12}C$-exchange, and some percent are chain elongation (Figure 3). The same should hold for fatty acids.

Text books of radiochemistry and hot atom chemistry teach us that there are also fragmentation products following the attack of the hot atom. These may be seen in our chromatograms (see Figure 2), but they have not yet been analyzed completely.

It is interesting to ascertain where the hot atom has replaced the stable carbon atom in the amino acid molecule. This has been done conclusively with the amino acid methionine. Amino acids react with ninhydrin by splitting off the amino and the carboxyl group and forming an aldehyde at the alpha position of the amino acid. The cleaved carbon dioxide can be easily measured as bariumcarbonate. Furthermore the S-methyl group may be separated from the methionine. The whole chemical scheme is to be seen in Figure 4. The relative numbers for the labeling sites are also given. It seems that the atoms of the endgroups are preferentially replaced. The amount of gaseous and volatile activation products is given in Figure 5.

If one irradiates substances containing fluorine-19 the liberated fluorine atom forms new bonds besides undergoing replacement (retention). Due to the high reactivity of fluorine,

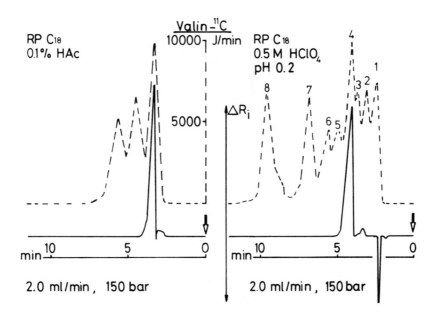

FIGURE 2. Activation of the amino acid valine. High pressure liquid chromatography in two systems like indicated. The third fraction (left part) and fraction 8 (right part) are eluted at the volume or time of the amino acid leucine.

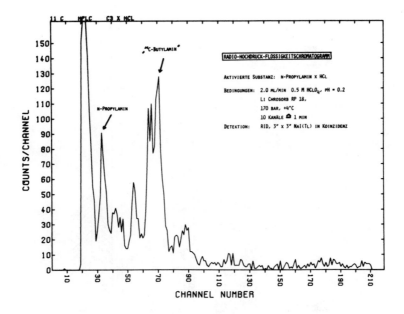

FIGURE 3. Activation of n-propylamine. High pressure liquid chromatography as indicated. At the volume of the butylamines there is to be found a radioactivity peak.

fluorination should not only occur in the hot atom state. So, by using monofluoroacetic acid as target,[9] one gets ^{18}F-labeled monofluoroacetic acid besides some other ^{18}F-containing substances (Figure 6). If one mixes a substance containing fluorine with another substance without fluorine atoms then the latter becomes ^{18}F-labeled after irradiation (Figure 7). As already mentioned, this must not only be due to the hot fluorine atoms. Therefore one may expect some preferential labeling positions depending on the chemical characteristics of the labeled molecule.

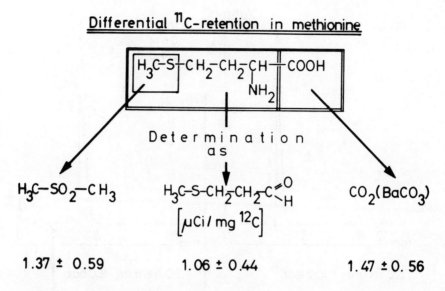

FIGURE 4. Activation of methionine. Activity distribution in several parts of the molecule.

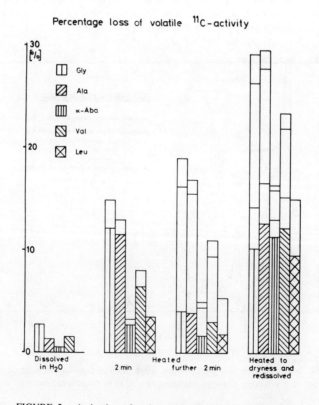

FIGURE 5. Activation of amino acids. Portion of gaseous and volatile products after different successive treatments as indicated.

It is possible to use Freon, a coolant, as fluorine donor. Acetic acid is readily mixed with Freon-11. After irradiation there is to be found a substance (Figure 7), which chromatographically behaves like monofluoroacetic acid.[10] Freeze-dried amino acid immersed in

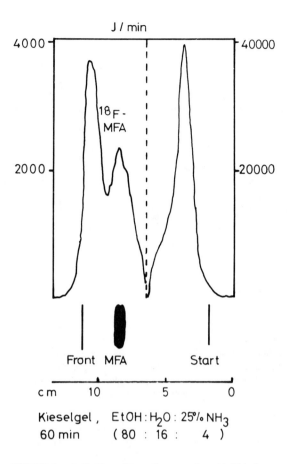

FIGURE 6. Activation of monofluoro acetic acid. Thin-layer chromatography as indicated.

Freon-11 leads to substances which electrophoretically no longer have the attributes of an amino acid (Figure 8). Thus by suitable separation methods it may be possible to isolate useful ^{18}F-labeled substances.

Some less efficient nuclear reactions of high energetic gamma quanta can be used for carrier-free radionuclide production. Irradiation of sodium yields radioactive fluorine by the ^{23}Na$(\gamma,n\alpha)^{18}$F-process. ^{28}Mg may be induced in silicon. Instead of silicon,[11] silica can be applied. In both reactions radioactive sodium is produced in abundance, but it is very easy to separate this quite well by chromatography.

It is also possible to get ^{123}I via the irradiation of Xenon. But this iodine isotope is always contaminated with ^{125}I. Therefore only the utilization of isotopically enriched ^{124}X could bring about good results.[12]

An electron linear accelerator will never be able to compete with cyclotrons with respect of radionuclide production. But one should see its chance. If an electron linear accelerator is present it can be utilized successfully. There is, furthermore, no need for chemical synthesis, and because of this, no need for a great chemical stuff. Beyond this, there is a chance, seldom used, for special activation analysis via the γ,n-reaction or the γ,γ'-reaction.

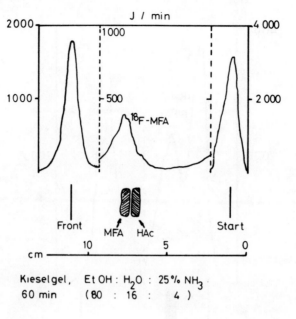

FIGURE 7. Activation of a mixture of acetic acid and monofluoro trichloro methane (Freon-11). Thin-layer chromatography as indicated: three peaks of radioactivity: one at the front correspondents with Freon, one is at the same place like monofluoro acetic acid and at the start there should be fluoride.

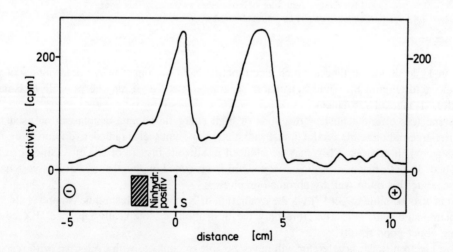

FIGURE 8. Activation of a mixture of alanine with Freon-11. The two peaks of activity migrate opposite to the amino acid.

REFERENCES

1. **Wolf, A. P.**, Labelling of organic compounds by recoil methods, *Annu. Rev. Nucl. Sci.*, 10, 259, 1960.
2. **Stöcklin, G.**, Chemie heißer Atome, Verlag Chemie, Weinheim, 1969.
3. **Theard, L. M. and Groce, D. E.**, Production of carbon-11 labeled compounds for nuclear medical application using an electron linear accelerator, in *Progress in Biomedical Engineering*, Fowel, L. J. and George, F. W., Eds., Sparta Books, Washington, 1967, 10.
4. **Loose-Wagenbach, I. and Clausnitzer, G.**, A system for production and transport of ^{11}C-Isotopes, *Nucl. Inst. Meth.*, 150, 345, 1978.
5. **Trampisch, W. and Sattler, E. L.**, ^{11}C-Rückstoßmarkierung organischer Moleküle für Biowissenschaften und Nuklear Medizin, *Kerntechnik*, 18, 222, 1976.
6. **Wagenbach, U., Gundlach, G., Trampisch, W., and Sattler, E. L.**, Direkte Markierung von Valin durch ^{12}C(γ,n)^{11}C-Kernumwandlung, *Naturwissenschaften*, 7, 340, 1967.
7. **Gundlach, G., Sattler, E. L., Trampisch, W. and Wagenbach, U.**, Untersuchungen zur direkten Markierung von Aminosäuren durch hochenergetische ^{11}C-Atome aus Kernreaktionen, *Z. Naturforsch.*, 31c, 377, 1976.
8. **Libby, W. F.**, Chemistry of energetic atoms produced by nuclear reactions, *J. Am. Chem. Soc.*, 69, 2523, 1947.
9. **Donnerhack, A. and Sattler, E. L.**, ^{18}F-recoil labelling with an electron linear accelerator, *Radiochem. Radioanal. Lett.*, 43, 393, 1980.
10. **Donnerhack, A. and Sattler, E. L.**, Production of carrier free ^{18}F-labeled acetic acid via a recoil labelling method, *Eur. J. Nucl. Med.*, 5, 277, 1980.
11. **Yagi, M., Yamadera, A., Fujikawa, S., and Shiokawa, T.**, Preparation of carrier-free ^{28}Mg by means of photonuclear reactions, *Int. J. Appl. Radiat. Isotopes*, 26, 637, 1975.

INDEX

A

A1 values for radionuclides, 35—36
A2 values for radionuclides, 35—36
Accelerator bombardment, 119
Acetylation, 83
Activation analysis, 165
Adrenal imaging, 104
Adsorption, 10—11, 27, 120, 130, 135
Aging of preparations, 11
Agricultural research, 107
Agricultural tracer use, 119
Air changes in radionuclide laboratory, 25
Air conditioning, 25
Air monitoring, 25
Aliphatic amines, 82
Aliphatic hydrocarbons, 81
Alkylation, 83
Alpha emitting impurities, 32
Alumina bed in technetium generators, 133—136
Aluminum, 138
Aluminum-28, 159
Aluminum nitride, 27
Amination, 81—82
Amino acid activation, 162—164
Angular correlation, 119
Anhydrous fluorine-18, 108—109
Anhydrous $H^{18}F$, 108
Annealing, 21
Annihilation, 48—49, 53
Anticancer chemotherapeutic, 83
Antitumor agents, 83
Aromatic amines, 82
Aromatic hydrocarbons, 81
Aromatic nitrils, 82
Arsenic-74, 118
Arsenic target, 113, 119
Astatine, 117—118
Asymmetric synthesis, 82
Automatic alarm systems, 25
Availability, see Commercial availability

B

$Ba^{14}CO_3$, 27
Barium-133m, 118
Benzodiazepine receptors, 83
Beryllium-7, 118
Beryllium nitride, 27
Beta decay process, 48—49
Beta-gamma ionization chambers, 25
Beta ray absorption measurements, 31—32
Beta ray scintillation spectrometry, 31
Billiard ball process, 161
Biological half-life, 6
Biological purity
   sublimed technetium-99m, 142
   technetium-99m generators, 138—139
Biological studies, 154, 160
Biology, 107, 110, 118—121 161
Biosynthetic preparation, 83
Bleeding of parent nuclides from generators, 11
"Bolus" administration in dynamic studies, 38
Bone marrow scintigraphy, 88
Bone scanning, 107
Brain metabolism study, 104
Brain opiate receptors, 83
Bromine, 104
Bromine-75, 111—114
Bromine-76, 111—114
Bromine-77, 111—114
Bromine-82, 113
Burn-up of target, 21

C

$^{44}CaCO_3$, 27
Calcium-45, 27—28
Cancer suppressing pharmaceutical, 104
Cancer therapy, 154
Carbon-11
   decay, 55
   nuclear production reactions, 55—57
   production, 55—65
Carbon-11-labeled acetylene, 62
Carbon-11-labeled carbon dioxide, 57—59
Carbon-11-labeled carbon monoxide, 59—60
Carbon-11-labeled cyanide, 60—62
Carbon-11-labeled formaldehyde, 64—65
Carbon-11-labeled methyl iodide, 62—63
Carbon-11-labeled methyllithium, 63—64
Carbon-11-labeled organic compounds, 80—83
Carbon-11-labeled oxides, 57—60
Carbon-11-labeled phosgene, 65
$^{12}C(d,n)^{13}N$ reaction on carbon dioxide, 72
$^{12}C(d,n)^{13}N$ reaction on solid targets, 70—72
$^{14}CH_4$, 27
$^{14}CO_2$, 27
Carbon-14, 11
   large-scale production, 27
Caroxylation, 81
Carrier Added, 51
Carrier contamination, 52
Carrier-free, 10, 14—15, 51—52, 58
   fluorine-18, 109
Carrier-freeness, 51, 58
Cation exchange separation, 28
Centralized radiopharmacy, 149
Cesium-134, 6, 11
Cesium-137, 18—19
Chain elongation, 161—162
Charged particle reactions, 51
Chemical purity, 6, 9
   evaluation, 32

solvent-extracted technetium-99m, 145
sublimed technetium-99m, 142
technetium-99m generators, 137—138
Chemicals for production of radionuclides, 24—25
Chlorine, 104
Chlorine-34m, 110, 158
Chlorine-38, 20
Chromatographic generators of technetium-99m, 133, 145—146
Chromatography, 165
Chromium-48, 118
Chromium-51, 8, 11, 20
large-scale production, 28—29
Chromium target, 120
Class I radionuclide laboratory, 25
Cobalt-55, 118
Cobalt-57, 118
Cobalt-58
large-scale production, 29
Ni metal for production, 24
Colorimetric methods, 32
Collimator, 53, 127
Column chromatography, 32
Commercial availability
radiohalogens, 104
reactor-produced radionuclides, 10—13
useful cyclotron nuclides with middle and relatively long lives, 118
Compact cyclotrons, 38
Computerized tomographic systems, 50
Concentration of additives, 32
Concentration of vehicles, 32
Containers, 24—25
Control instrumentation, 25
Control of radiation hazards, 26
Control samples, 26
Copper-67, 118
Coprecipitation, 113
"Critical organs," 5
Cross-sections, 11, 14—15, 18, 21—23, 55—57, 66, 71—72, 74—76, 129, 155
Current availability of reactor-produced radionuclides, 35
Cyclocondensations, 81
Cyclotron, 35, 51
Cyclotron nuclides with middle and relatively long lives, 118—121
Cyclotron radionuclides, see also specific radionuclides, 35, 38, 103—124

D

Decay
by positron emission, 48—49
carbon-11, 55
nitrogen-13, 65
oxygen-14, 74
oxygen-15, 74
Decay constant, 21, 24
Delay tanks, 26

Demethylation, 82
Diagnosis, radionuclides application in, 49—50, 126, 154
Diagnostic tracer studies, 5
Direct fluorine-18 labeling of organic compounds, 109—110
Disposal of wastes, 26
Distillation, 113, 118
Dopamine receptors, 82—83
Dry distillation method, 27, 31
Dry vaporization, 118

E

Effective half-life, 50
Effluent storage, 26
Electromagnetically enriched $^{44}CaCO_3$, 27
Electromagnetically enriched ^{50}Cr, 28
Electromagnetically enriched ^{58}Fe, 29
Electron, 48—49
Electron linear accelerator, 165
labeled compounds produced at, 161—167
radionuclides produced at, 161—167
Electronic excitation, 51
Electronic interactions, 51
Elution efficiency of a generator, 133—135
Elution of technetium-99m, 133—134
Emission spectrography, 32
End of bombardment (EOB), 51, 81
Enriched target, 105, 119, 129
Enrichment factor, 21
Environmental contamination control, 26
Enzymatic amino acid preparation methods, 82
Enzymatic dephosphorylation, 83
Enzymatic methods, 83
advantage of, 84
Erythrocytes, technetium-99m labeling of, 138
Erythropoietic function, 88
Estrogen receptors, 83
Excitation functions, 105, 110—113, 115—120
Exothermic fusion reaction, 154

F

Fatty acid synthesis, 81
Fissile alpha-emitters, 137
Fission, 13, 18—19
Fission product impurities, 32
Fission product radionuclides, 18—19
Fission yields, 19
Fluorinating agent, 108
Fluorine, 104
$^{18}F_2$ production, 109
Fluorine-18, 24
anhydrous, 108—109
direct labeling of organic compounds, 109—110
14 MeV neutron generator production of, 158
no carrier added, 105—108
nuclear production reactions, 105

production, 104—110
production reactions, 104—105
radiochemical purity, 158
Fluorine-18-labeled 2-deoxy-2-fluoroglucose, 109
14 MeV neutron generators, see also specific radionuclides, 154—155
  preparation processes, 158—160
  production of radionuclides, 155—157
  separation methods, 156, 158
Fragmentation products, 162
Free radical species, 133

G

Gaiters, 25
Gallium-67, 38, 118—119
(γ,n)-reaction, 161
Gamma-ray emitters, 136
Gamma spectrometry, 31—32
Gas chromatography, 27, 52
Ge(Li) crystal detectors, 31—32, 136
Generator concept, 127
Generators, see also specific types, 133
  short-lived positron emitters, 84—85
^{68}Ge-^{68}Ga system, 84—85
Germanium target, 119
Giant resonance, 161
Glass containers, 25
Glucose metabolism, 83
Gold-198, 20
Gold-199, 20
Gravimetric methods, 32

H

H_2O_2 in irradiated MoO_3, 11, 13
Halogen labeling, 104
Halogen substitution, 104
Handling facilities, 25—26
Heart diagnosis, 119—120
Heart studies, 89
Heavy metal contaminants, 9
High energy spallation reactions, 121
High kinetic energy, 51
High pressure liquid chromatography, 52, 67
High specific activity, see also Specific activity, 9—10
  molybdenum-99, 29
Historical survey of developments in tracer application, 4
Hold-back carrier, 10, 32
Hormone sensitive tumors, 83
Hot atom reactions, 51—52, 77
Hot atoms, 51, 72, 161—163
Hot cells, 25
Hydrogen-3, 11
Hydrogenation, 82

I

Immobilization of long-lived wastes, 26
Immobilized enzymes, 84
Impurities, 52, 135, 137
Incineration of wastes, 26
Indirect reactions, 20
Indium-111, 118—119
In-process control equipment, 26
Internal radiation source, 118
Iodine, 104
Iodine-123, see Radioiodine
Iodine-123/Iodine-124 activity ratio, 114
Iodine-124, 114, 116
Iodine-125, 11, 20, 114, 117
  labeling of proteins and hormones with, 9
  large-scale production, 30
  protein iodination, 11
  reducing agents in, 11
  specific activity of, 10
Iodine-126, 20
Iodine-126 in Iodine-125, 30
Iodine-127, 16
Iodine-128, 14 MeV neutron generator production of, 159
Iodine-129, 16
Iodine-131, 5, 11, 114, 117
  chemical impurities, 9
  dry distillation method to produce, 31
  growth of microorganisms in, 13
  labeling of proteins and hormones with, 9
  large-scale production, 30—31
  oxidation, 31
  protein iodination, 11
  radiochemical impurities, 8
  reducing agents in, 11
  self-radiolysis in solutions of, 10, 31
  specific activity, 9—10, 18
  Te metal or TeO_2 for production of, 24
  tellurium target irradiation to produce, 15—16, 18
  wet distillation method to produce, 30—31
Iodine target, 116
Ion exchange, 119, 121
Ionization, 51
Ionization chamber, 26
^{52}Fe-$^{52}\Sigma$Mn system, 84
Iron-52, 87—88, 118
$^{58}Fe_2O_2$, 24
Iron-59, high specific activity
  $^{58}Fe_2O_3$ for production of, 24
Iron-59, large-scale production of, 29
Iron-82, 120
Irradiation, TeO_2 for, 24
Irradiation facilities, 24
Isotopic exchange, 114

K

Kidney studies, 89

Krypton-76, 113
Krypton-77, 113
Kryptons, 111—112, 114

L

Labeled compounds
  diagnostic use of, 50
  electron linear accelerator, production at, 161—167
  preparation of, 8—9
  special requirements for, 11—13
Large-scale producers supplying reactor-produced radionuclides, 35, 37—38
Large-scale production considerations, 24—26
Leaching, 11, 30
Lead-203, 118, 121
Life sciences, 120
Linear accelerators, 51
Lithium-6, 20
Long-lived radionuclides, 5
Long-term tracer experiments, 111
Lung ventilation studies, 85—86

M

Magnesium-27, 14 MeV neutron generator production of, 159
Magnesium-28, 118, 120—121
Main production reactions, 13—21
Maintenance of remote processing facilities, 26
Manganese-52, 118
Manganese-54, 118
Manganese-56, 159—160
Marketing set up, 26
Master-slave manipulator, 25
Medical diagnosis, see Diagnosis
Medical imaging, 53—54
Medicine, see also Nuclear medicine, 118—121, 161
  radionuclides application in, 49—50
Mercury-203, 5
Metabolic studies, 11, 160—161
Metal impurities, 32
Metallurgy, 120
Methylation, 82—83
Methyl ethyl ketone, 143, 145
Microbiological attack, 10
Microorganisms, growth of, 13
Microtitration, 32
Middle and relatively long-lived cyclotron radionuclides used in biology and medicine, 118—121
Mini manipulator, 25
Molybedum, ^{99}Mo production through neutron activation of, 129—130, 132—133
Molybdenum metal, 130
Molybdenum-99, see also Technetium-99m, 127
  direct irradiation of molybdenum oxide to produce, 29
  fission of uranium to produce, 29—30, 130—133
    quality control of, 11
  high specific activity, 29
  large-scale production, 29—30
  methods of separating technetium-99m from, 133—139
  neutron activation of molybdenum to produce, 129—130, 132—133
  production of, 129—133
    choosing method of, 132—133
  specific activity of, 10—11
Molybdenum-99:Technetium-99m generators, see also Technetium 99m, 127
  kinetics of growth and decay of, 127—130
Molybdenum trioxide, 130, 139, 141—142
Mössbauer spectroscopy, 119
Multichannel analyzer, 136
Multisection imaging devices, 54
Myocardial imaging, 87, 89

N

(n,α) reactions, 13, 16, 18—19, 23, 155—156, 158
(n,γ) followed by decay, 13—18
(n,γ) reactions, 13—14
(n,p) reactions, 13, 16, 18—19, 23, 155—156, 158
$(n,2n)$ reactions, 155, 157—158
NaI(Tl) crystal detectors, 31, 127
Neon-19, 85—86
Neutrino, 48—49
Neutron activation of molybdenum, 129—130, 132—133
Neutron flux, 21—24, 129—130
Nickel-57, 118
Nickel metal, 24
Nitrogen-13
  decay, 65
  nuclear production reactions, 65—66
  production of, 65—74
Nitrogen-13-labeled ammonia, 67—69
Nitrogen-13-labeled molecular nitrogen, 70—73
  $^{12}C(d,n)^{13}N$ reaction
    on carbon dioxide, 72
    on solid targets, 70—72
  (p,α) reaction on water, 73
  (p,n) reaction on enriched carbon-13, 72—73
Nitrogen-13-labeled nitrates, 66—67
Nitrogen-13-labeled nitrites, 66—67
Nitrogen-13-labeled nitrogen oxides, 74
Nitrogen-13-labeled organic compounds, 80—81, 83—84
Nitorgen dioxide production, 74
No Carrier Added, 51, 119
  fluorine-18, 105—108
  H^{18}F, 108
Nuclear medicine, see also Medicine, 38, 104, 114, 119, 121

technettum-99m, position of, see also Technetium-99m, 125—152
Nuclear production reactions
   carbon-11, 55—57
   fluorine-18, 105
   nitrogen-13, 65—66
   oxygen-14, 74—76
   oxygen-15, 74—76
   useful cyclotron nuclides with middle and relative long lives, 118
Nuclear reaction, 161
Nucleus, 48—49
Nuclide generator, 20
   system, 50

O

Operations and maintenance cell, 26
Organ imaging, 147
   techniques, 126—127
Organic chemistry, 110
Organic compounds
   carbon-11 labeling, 80—83
   direct fluorine-18 labeling, 109—110
   nitrogen-13 labeling, 80—81, 83—84
Organic fluorine compounds, 108
Oxidation, 131
Oxidizing agents, 138
Oxygen-14
   decay, 74
   nuclear production reactions, 74—76
   production, 74—80
Oxygen-14-labeled molecular oxygen, 79—80
Oxygen-14-labeled water, 80
Oxygen-15
   decay, 74
   nuclear production reactions, 74—76
   production, 74—80
Oxygen-15-labeled carbon dioxide, 78
Oxygen-15-labeled carbon monoxide, 78
Oxygen-15-labeled molecular oxygen, 76—77
Oxygen-15-labeled water, 79

P

(p,α) reaction on water, 73
(p,n) reaction on enriched carbon-13, 72—73
Packing, 32—35
   materials, 24—25
Palladium membrane, 27
Pancreas diagnosis, 119
Paper chromatography, 32
Paper electrophoresis, 32
Parent nuclides, 11
Personnel exposure, 26
Pertechnetate, 137—138, 142
pH meters, 25—26
Pharmacokinetic investigations, 83
Phosphorus-30, 86

Phosphorus-32, 11
   adsorption of, 10
   growth of microorganisms in, 13
   large-scale production, 28
   microbiological decomposition, 10
   sulfur for production of, 24
Photosynthetic preparation methods, 83
Plutonium-237, 118, 121
Plutonium-239, 18
Pneumatic transfer facilities, 24
Polyethylene containers, 25
Positron, 48—49
Positron emission
   decay by, 48—49
   tomography, 53
Positron Emission Transaxial Tomograph (PETT), 53—54
Positron emitters, 38
Positron emitting radionuclides, see also specific radionuclides, 47—101
Positron imaging devices, 53—54
   multisection, 54
   single section, 53
Positronium atom, 49
Potassium-38, 86—87
Potassium-42, large-scale production of, 27—28
Potassium-43, 118, 120—121
Potassium fertilizers, 120
Practicable specific activities, 10
Practical half-life, 3—5
Precipitation, 28
Precursor of fluorine-18 labeling 108—109
Preservatives, 9—10
Price of reactor-produced radionuclides, 10—13
Primary containers, 32
   glass or polyethylene, 25
Processing facilities, 25—26
Product radionuclides, 14—15, 18, 21
Production of radionuclides, see specific topics
Production rates of Iodine-123 and Iodine-124, 116
Production reactions, see Nuclear production reactions
Protective equipment, 26
Protein iodination, 11
Proton irradiation, 81
Psychoactive drugs, 82—83
Pulmonary investigations, 72
Pyrophoric U, 27

Q

Quality controls, 19, 31—32
Quantification, 53—54

R

Radiation dose
   contribution of impurities, 6
   thyroid diagnosis, 115

Radiation exposures, 5, 126—127
Radiation sources, 119
Radiation therapy, 118
Radioactive decay law, 127
Radioactive wastes, 11
Radioassays, 38
Radiobromine, 111—114
    chemical processing, 113—114
    14 MeV neutron generator production of, 158—159
    labeling, 114
    nuclear production reactions, 111—112
    target processing, 113—114
    useful nuclides, 111
Radiochemical fume hoods, 25
Radiochemical purity, 6, 8—9
    aluminum-28, 159
    chlorine-34m, 158
    evaluation, 32—33
    fluorine-18, 158
    manganese-56, 159—160
    technetium-99m generators, 137
Radio-frequency fusion, 119
Radio-frequency heating, 113
Radiohalogens, 104—118
    astatine, 117—118
    chlorine-34m, 110
    commercial availability, 104
    fluorine-18, 104—110
    radiobromine, 111—114
    radioiodine, 114—117
    use of, 104
Radioimmunoassays, 3, 10—11, 114
Radioiodine, 38
    labeling, 117
    neutron-deficient, 114
    production
        direct route, 115
        rates, 116
        reactions, 114—115
        via xenon-123, 116—117
    use of, 117
Radiokryptons, 111, 114
Radiolysis, 51, 58
Radiolytic decomposition, 52
Radionuclides, see specific topics
Radionuclidic impurities, 14
Radionuclidic purity, 6—8, 116
    evaluation, 31—32
    solvent-extracted technetium-99m, 144—145
    sublimed technetium-99m, 140—142
    technetium generators, 135—137
Radiopharmaceuticals, 49—50, 138
    preparation of, 11—13
    technetium-99m, 147—149
Radiotracers, 8—9
Radiothalliums, neutron-deficient, 119
Radioxenons, 114, 116
Rapid turn-over studies, 87
Raw materials including targets, 24—25
RBC labeling, 8

Reactor-produced radionuclides, 1—45
    calculation of yields and specific activities, 21—24
    current availability, 35
    future perspectives, 35, 38
    large-scale production considerations, 24—25
    main production reactions, 13—21
    packaging, 32—35
    quality controls, 31—32
    tracers, general requirements for use as, 3—13
    transport, 32—35
    typical production processes, 26—31
Recoil chemical form, 14, 20
Recoil energy, 51, 161
Recoil enrichment recovery, 20
Recoil labeling, 161
Recoil reaction, 52
Reduced pressure distillation, 28
Reducing agent impurities, 32
Reducing radical species, 133—134
Reductive methylation, 82—83
Remote handling facilities, 24—25
Remote manipulation, 24—25
Remote pipetting units, 25
Retention, 20—21, 161—162
Rhenium, 139, 141, 143
Rhenium-188, 20
Rhodium-101m, 118
Rubidium-81, 88—89
Rubidium-86, large-scale production of, 27—28
Ruthenium-97, 118
Ruthenium-103, 11

S

Safety interlocks, 25
Sales set up, 26
Saturation-specific activity, 21—22
Saturation value, 21—22
Scattering phenomena, 51
Scavenging agents, 138
Scintigraphic techniques, 50
Scintillation camera, 53
Sealed-tube neutron generators, 154
Selective adsorption, 28
Selenium-72, 120
Selenium-73, 118—120
Selenium-75, 11, 120
Selenium target, 113
Self-radiolysis, 10—11, 31
Servicing of remote processing facilities, 26
Shielded vaults for storage of active solid wastes, 26
Shielding container, 32
Shippers certificates, 35
Shipping containers, basic requirements for, 32
Shipping documents, 35
Short-lived labeled compounds, preparation of, 51—53
Short-lived positron emitters

application of, 49—50
generators for, 84—85
Short-lived positron emitting radionuclides, see also specific radionuclides, 47—101
Short-lived radionuclides, 5, 24
  defined, 48
  14 MeV neutron generator production of, 153—160
Single section imaging devices, 53
Sodium-22, 118
Sodium-24, large-scale production of, 27
Solidification of long-lived wastes, 26
Solvent extraction, 118—119
Solvent extraction generator, 142—146
  chemical purity, 145
  efficiency of, 144
  radionuclide purity, 144—145
Special form radioactive material, 35
Specific activity, 5—6, 9—11, 13—16, 18, 20—23, 27—31, 50—51, 56, 109, 120, 129
  calculation of, 21—24
  estimation of, 32
Specific iodide electrode, 32
Spot tests, 32
Sterility, 139
Steroidal hormones, 38
Storage of effluent and waste, 26
^{82}Sr-^{82}Rb system, 85
Strontium-90, 11, 18—19
Sublimation generator, 130, 139—142, 146
  biological purity, 142
  chemical purity, 142
  other purity aspects, 142
  radionuclidic purity, 140—142
Successive neutron capture, 23
  reactions, 13, 20
Sulfur, 24
Sulfur-35, 11
  large-scale production of, 28
Sulfur-38, 20
Sump tanks, 26
Sweep gas, 76
Synthesis, 52
Szilard-Chalmers recoil enrichment, 13—14, 20—22, 28—29, 130, 158, 161

T

Target assembly, 105, 108—109
Target gas, 109
Target no-carrier-added fluorine-18, 108
Target production of ^{18}F-F$_2$, 110
Targetry, 112—113
Targets, 10, 14—16, 18—19, 22, 24
  irradiation in nuclear reactor, 21
  raw materials including, 24—25
Technetium, 126
  chemistry, 147—148
  radiopharmaceuticals, 147—149
Technetium generator, see Technetium-99m

Technetium heptoxide, 139
Technetium-99m, 5—6, 11
  alumina bed in generators, 133—136
  biological purity, 138—139, 142
  chemical purity, 137—138 142, 145
  chromatographic generators, 133, 145—146
  comparison of generator forms, 145—146
  concentration profile, 133
  elution efficiency, 133—135
  elution profiles, 133—134
  generator concept, 127
  generators, 133
  impurities, 135, 137
  isomeric level decay, 127
  methods of separating from molybdenum-99, 133—139
  physical properties, 126
  radionuclidic purity, 135—137, 140—142, 144—145
  radiopharmaceuticals of, 147—149
  solvent extraction generator, 142—146
  special position in nuclear medicine, 125—152
  sterility, 139
  sublimation generator, 139—142, 146
  suitability for organ imaging, 126—127
  supply considerations, 127
Tellurium metal, 24
TeO$_2$, 24
Thallium-201, 118—119
Thulium-167, 118
Therapy
  cancer, 154
  radiation, 118
  radionuclides application in, 49
Thermal diffusion process, 27
Thermal reaction, 52
Thick target saturation activities
  bromine, 113
  fluorine-18, 106—107
  selenium, 120
Thick target yield, 105—107, 110—111, 116, 119
Thin-layer chromatography, 32
Threshold energy, 22
Thyroid diagnosis, 114
  radiation dose, 115
Thyroid gland, 104
Thyroid studies, 8—9
Tin-113, specific activity of, 11
Titration, 32
Tomographic imaging system, 53
Toxicity, 5—6
Tracers
  biomedical sciences and, 2—3
    criteria and requirements, 3—13
    physical data for, 41—43
  radionuclides, applications of, 3
  reactor-produced radionuclides
    applications, 39—40
    requirements, 3—13
  study, 104—110
Transamination, 84

Transfer ports, 25
Transport-Index, 34
Transport regulations, 32—35
Treatment of wastes, 26
Tritium, large-scale production of, 27
Tungsten-188, 20
Type A containers, 34—35
Type B containers, 34—35
Type B(M) packages, 35
Type B(U) packages, 35
Typical production processes, 26—31

U

Ultra short-lived radionuclides, 3, 5
Unperturbed neutron flux, 22
Uranium-325, 11, 18
Uranium fission, molybdenum-99 produced through, 130—133
Useful cyclotron nuclides with middle and relatively long lives, 118—121

V

Vanadium-48, 118
Van de Graaff generators, 51, 66, 72, 76
Ventilation system, 25

Vial sealing machines, 25—26
Volatilization, 113, 119
Volumetric methods, 32

W

Waste collection, 26
Waste treatment, 26
Wet distillation method, 27, 30—31

X

^{122}Xe-^{122}I system, 85
Xenon-123, 116—117
Xenon-124, 30
Xenon-125, 117

Y

Yields, calculation of, 21—24, 75

Z

Zinc-62, 118
^{62}Zn-^{62}Cu system, 84